THE UNITY OF TRUTH

The UNITY of TRUTH

Solving the Paradox of Science and Religion

ALLEN A. SWEET,
C. FRANCES SWEET,
AND
FRITZ JAENSCH

iUniverse, Inc.
Bloomington

THE UNITY OF TRUTH
SOLVING THE PARADOX OF SCIENCE AND RELIGION

iUniverse books may be ordered through booksellers or by contacting:

iUniverse
1663 Liberty Drive
Bloomington, IN 47403
www.iuniverse.com
1-800-Authors (1-800-288-4677)

ISBN: 978-1-4759-3060-3 (sc)
ISBN: 978-1-4759-3059-7 (hc)
ISBN: 978-1-4759-3058-0 (ebk)

Library of Congress Control Number: 2012910094

Printed in the United States of America

iUniverse rev. date: 09/08/2012

CONTENTS

TESTIMONIALS

"The Unity of Truth" is a commendable attempt in seeking the unification of the scientific truth and the religious truth— the greatest riddle before human kind. The exposition is lucid, methodological and logical. The line of argument teaches us as it delights. This book will undeniably constitute an irrefutable and convincing approach in establishing the unity of truth as inferred by votaries of science and that perceived by followers of religion. Any rational mind familiar with scientific methodology is bound to appreciate the conclusions arrived at, as all his articles of faith remain intact. The main plank on which the argument rests is in invoking the a-causal behavior of quantum mechanics at measurement as God's message to human beings without violating the laws of conservation of energy.

Dr. S. P. Puri, Professor and past Chariman, Department of Physics, Panjab University, Chandigah India

A rich portion of the text is concerned with paradox. As I read, I realized that scientists and mystics often take a very different approach. Whereas scientists seek to reconcile paradoxes, mystics tend to enter them and pray them. Therefore, to this mystic, quantum activity becomes a way to analogize the divine rather than explain the divine. Part of my approach to God, as you say elsewhere in the text, is that I can never know God as God is, but I can hear those parts of God that speak to me. Quantum theory helps to give form to this experience.

Rev. Carol Luther, Episcopal Priest and Chaplain of St. Paul's School, Oakland California

Thank you for allowing me to read and review your book. I am now thinking about my spiritual and earthly journey in ways that I have not previously considered as a result of this experience. What a gift!

Connie Conley-Jung Ph.D, Clinical Psychologist in private practice, Alameda California

This book provides the basis for a dialogue bringing science and religion into a single cohesive unit.

Gill Lane MS, Analog Electronics Consultant in private practice, Palo Alto California

This book is dedicated to all who have felt, at some moment in their lives, that they must check their brain at the door of their church and/or felt that they must keep their religious faith to themselves so their scientific colleagues would not label them brainless, or worse. Wholeness is absolutely essential if we are to achieve any kind of satisfaction in our lives. In the highest sense, this book is about the quest for personal wholeness and for regaining a consensus of trustworthy authority to guide us forward.

Well, I do not think that it is necessarily the case that science and religion are natural opposites. In fact, I think that there is a very close connection between the two. Further, I think that science without religion is lame and, conversely, that religion without science is blind. Both are important and should work hand-in-hand. It seems to me that whoever doesn't wonder about the truth in religion and in science might as well be dead.

—Albert Einstein[1]

---

[1] P. Bucky, A. Weakland, and A. Einstein, *The Private Albert Einstein* (Riverside, NJ: Andrews and McMeel, 1992), 85.

PREFACE

About the Authors

The work of a group is always the product of the unique backgrounds and talents of its members. This statement is so true with this book. Each of us has brought his (and her) own unique insights, talents, and perspectives to bear in ways that could never have been foreseen before starting the work. We, the authors, now want to share with our readers a little information about each of our backgrounds and personal history.

Allen Sweet has been fascinated with science almost since birth. He was born in Providence, Rhode Island, during WWII. His first scientific experiment was conducted at age three, when he knocked over a table lamp, breaking its bulb. Then he realized that this was his chance to find out if the glass surrounding a light bulb was really necessary for the production of light. Allen turned the light switch on, and the bulb's filament blazed brightly and burned out. Allen had the answer to his first experiment. Unfortunately, the broken bulb had cut his hands. When his mom came into the room and saw the broken glass and his bloody fingers, she put a stop to Allen's experiments for some time to come.

Not to be deterred, Allen dabbled in kitchen chemistry, fossil hunting, and model airplane building for a number of years. However, he found his true love in the form of a crystal radio set he built at age ten. By age thirteen, he was a licensed ham radio operator, routinely talking to other ham operators around the world on radios he had built himself. This was at a time (the mid-1950s) when getting a call from two hundred miles away was considered an exciting long-distance phone call.

At this time, a few hams with budgets considerably larger than a teenager's like Allen were experimenting with bouncing their signals off the moon. This kind of transmission required the use of a part of the radio spectrum called microwaves. Reports of these "moon-bounce" experiments inflamed Allen's imagination, and he longed to know more about how to accomplish such amazing feats. So great was his curiosity that Allen and several of his young friends visited the home of one of the leading moon bouncers completely unannounced. As luck would have it, they were treated to a grand tour of the station. Even way back then, Allen was ready to devote his life to the science and technology of microwave communications.

After high school graduation, Allen enrolled in Worcester Polytechnic Institute to study electrical engineering. His first two years were primarily devoted to the study of physics, chemistry, and mathematics. Although Allen had been very active in his church while growing up, he no longer could relate to the institutional church because at this time, his most intense experiences of God seemed to come out of his exploration of the laws of science. Allen might have left the church, but he never left God.

After college graduation, Allen followed his dreams by enrolling as a graduate student in a multidisciplinary program in microwave science and technology at Cornell University. The faculty and students in this program came from the electrical engineering, physics, applied physics, and material science departments. Four years later, Allen graduated with MS and PhD degrees, ready to embark on a research career in microwave communications.

Over the years, Allen has published widely in technical journals and at conferences. He holds several patents and has written two popular reference books in the microwave communications field. In 1979 he received his field's highest award—the Microwave Prize—which is awarded each year by the Microwave Theory and Technique Society of the Institute of Electronics and Electrical Engineers. Today Allen operates a successful design and consulting company and is an adjunct professor at Santa Clara University. Allen's work has made many significant contributions to the worldwide revolution in wireless communications.

Sometime in midlife, Allen began to feel a call to search for a kind of knowledge that is beyond science. Together with his wife, Fran (also a coauthor of this book), he began an exploration of the world's religious traditions. In time this exploration led both Allen and Fran back to their Christian roots but with a spiritual knowledge that was greatly enriched by their experiences with the traditions of many cultures and religions. Both Allen and Fran became fascinated by the many striking parallels between the teachings of the world's religions and the laws of science. For many years they have been searching and discussing this topic endlessly. This book is the final culmination of their efforts.

C. Frances (Fran) Sweet has been deeply devout since childhood. Many of her earliest memories are about how much she was spiritually moved by the services and activities at her home church in Alameda, California. As she grew older, the nuns who taught her catechism class imparted to her a lifelong sense of God's presence in her life and the knowledge of her own spirituality. This sense of godliness has always been a very important part of who Fran is. At one point in her life, Fran seriously considered becoming a nun. However, since she was an only child, Fran didn't feel this path would be fair to her parents, who so looked forward to grandchildren.

Fran graduated from high school and went on to study the social sciences (with an emphasis in gerontology) at San Francisco State University, where she received a BA degree. For complex personal reasons, Fran left behind the institutional church of her youth and, like Allen, became a spiritual seeker. Allen and Fran met by chance when he answered a newspaper ad Fran had placed for her secretarial and word-processing service company. Both Fran and Allen were newly single, and in time, their relationship evolved from professional to personal. Within three years, they were married. Together they have seven grown children and one grandchild.

As Fran and Allen's spiritual searching led them back to their Christian roots, Fran began to feel a call to ordained ministry. She enrolled in the Episcopal Diocese of California's school for deacons. Allen joined her, and they both graduated three years later with a bachelor's degree in theological studies. One year later, Fran was ordained as a permanent deacon in the Episcopal Church. Since then her ministry has been roughly divided between parish ministry and hospital ministry. In 2011 she celebrated the twentieth anniversary of her ordination.

Allen, Fran, and their friend and coauthor Fritz have been discussing the parallels and contradictions of science and religion for the past six years. They meet every Saturday night at a local Starbucks, spending hours in high-spirited discussions of the topic of the moment. Both Fran and Fritz are poets. Here is a poem by Fran that perfectly captures the spirit of this book and the process by which it has been created:

Growing stronger, living longer.
Day by day, reaching farther,
In your very own way,
Very soon now, you'll have reached your dream,
You'll have met your goal,
Things are not always as they seem,
Sometimes they are better,
And sometimes—you will find a new dream!
Never stop dreaming—ever,
And never forget—
From your dreams great discoveries have grown,
Never forget—
It's never over until it's done!
So stand up! Put your hopes and dreams to work!
For such people, there is no stopping
There is no ending, but—
Always, there is always—
Another hope, another chance, another direction, another corner to
turn, another path to walk.
From great dreams such as yours,
Great things grow!
So stand up!
Put your feet on the floor,
Put your dreams to work!
There is always another chance, another choice,
another corner to turn!
Don't ever stop dreaming,
It's never over until it's done!
So!
Stand up! Put your dreams to work!
Because there is no ending, only beginning,
From such "dreams" great things grow.
You're on your way!
Hold on tight,
And—never let go!

Fritz Jaensch was born, grew up, and came to young adulthood during the amazing pre-and post-World War II period in Germany. Fritz was an eyewitness to one of the most confounding, troubling, and confusing periods in all of human history. Fritz can recall, through the eyes of a child, watching soldiers herd concentration camp victims through the streets of his town. His own mother spent part of the war in prison because she had helped a Jewish friend.

Fritz recalls that sometime during the period of hunger and devastation that followed the war, he found a church where a service was being held. The light streaming out of the church's windows and the joyful voices singing hymns inside called to Fritz to enter. He went inside, found God, and has never left him.

When Fritz was a young man, he immigrated to the United States, where he settled in the farm country of Northern California. Since farming was the work he knew best, Fritz went to work as a farmhand on a dairy. However, the people of his church in California thought that Fritz was destined for greater things and urged him to pursue an education. In time Fritz received a BA in history and German language and literature and an MA in history, both from Sonoma State University. He then enrolled in a PhD program in European history at UC Berkeley. Fritz has done a number of translations for the University of Alaska, Fairbanks, from the German language based on the writings of several of the early Russian explorers to Alaska and Siberia, one of these was published (*Siberia and Northwestern America, 1788–1792: The Journal of Karl Heinrich Merck*, Kingston, Ontario, Canada).

Together with his wife, Stella, Fritz founded a window-cleaning and janitorial company in Alameda, California. He continues to operate his business in partnership with his stepdaughter, Roxanne Green. Fritz is a historian, a Christian, and, as such, a student of Holy Writ. In all ways Fritz views the world with the eye of an artist and hears the world with the ear of a poet.

Acknowledgments

The authors wish to gratefully acknowledge, thank, and recognize the many helpful suggestions, comments, and encouragements they have received from the following individuals during the preparation of this book: John Shelby Spong, Carol Luther, Paul Bailey, Connie Connley-Jung, Satya Puri, Ramesh Puri, Marc Andrus, Stephen Kosslyn, Doug Jensen, Laura Frank, Gill Lane, Roxanna G. Green and Jill Sweet Dutterer.

Our special thanks go to the Rijksmuseum in Amsterdam, the Netherlands, for making Honore Daumier's painting "Jesus Christ and His Disciples" available for our book's cover art.

INTRODUCTION

Truth

Truth is such a beautiful word. It echoes timelessly with a meaning of something you can always believe in, even as the storms of controversy are swirling all around you. But whose truth do you believe? Pontius Pilate famously asked, "What is truth?" This question has left Bible readers for the past two thousand years pondering whether Pilate was simply ignorant of the greater philosophical issues or was indeed acutely aware of the issues involved and speaking to the profoundness of the question.

Whatever his state of mind, his question is a good one. What indeed is truth? In the world of the twenty-first century, most people on planet Earth look to two sources for the highest, most fundamental truth: science and religion. But this situation raises many difficult questions: Is scientific truth the highest authority and in a conflict with religious truth? Is religious truth the highest authority and in a similar conflict with science? Are religious truth and scientific truth mutually exclusive, or is truth relative, with no absolutes? Is there only one truth, and do what we call scientific truth and religious truth exist simply as facets on the same gem of truth?

It is the position of the authors of this book that the latter description of truth is in fact the reality. We are convinced of the existence of a single, unified truth. Our purpose in writing this book is to present evidence in support of this position. We are also convinced that religious people should become more open to the scientific approach to truth. Without the reason and observational knowledge of science, religion is in danger of falling into a rigid fundamentalism. On the other hand, scientists as human beings need to find that sense of awe that religion offers. They also need to share in the ethical and moral teachings of religion to avoid falling into a kind of amoral materialism that is so characteristic of the more-vocal atheistic scientists. Albert Einstein once said, "Science without religion is lame, and conversely religion without science is blind."[2] It is our position that both science and religion are needed for personal wholeness, and one without the other represents an incomplete truth.

Dreams

Dreams have long been viewed by many cultural traditions as harbingers of new directions. Two such dreams moved Allen Sweet, our primary author, in the direction that has culminated in the writing of this book. The first of these big dreams occurred in September 1962, during Allen's first week of studies at Worcester Polytechnic Institute in Worcester, Massachusetts:

> It was the first day of classes, and I was heading out the front door of my dormitory to my first year of college classes in physics, chemistry, and mathematics. As I left I noticed, in the way of dreams, that everything had changed. The now-familiar campus had been transformed into a magnificent gothic cathedral. Because there was no place else to go, I entered the cathedral in search of my first class meetings. However, instead of classrooms and laboratories, I found a great nave leading to a high altar, which was surrounded by many smaller chapels arranged in a semicircular pattern around the high altar.
>
> As I walked forward both in disbelief and some disappointment, I began to notice the names of the chapels: physics, chemistry, biology, mathematics, electrical engineering, mechanical engineering, chemical engineering, and civil engineering. In my dream, all of my present and future classes had become chapels arranged about the cathedral's high altar. In a split second, it dawned on me that my education was being moved to the church of science. I felt a profound sense of confusion and a loss of the familiar. I had come to college to begin my formal education in science and technology, but this dream was taking me to an

---

[2] Ibid.

unfamiliar church that claimed to be a church of science. Upon awakening, I wondered what this development could possibly mean. I quickly forgot all about this dream. I did not remember it or grasp its significance until many years later.

Here is Allen's second and equally profound dream:

I had this dream many years ago. I think it was in the fall of 1990 while I was camping near Olema, California. The exact day of the dream was the eve of the feast of Christ the King. I found myself walking through an old English churchyard. As I walked among the ancient, weathered headstones, I noticed a statue in their center. The statue was of Christ the King seated on a throne, crowned with a crown of thorns, and holding a cup of wine in one hand and a loaf of bread in the other.

As I approached the statue, I could see the statue was very old, perhaps ancient. The statue was made from poured concrete that was composed of a coarse aggregate of sand, pebbles, and stones. Over the years, the statue had weathered and decayed to the point where its features were barely recognizable. I stood looking at the face for a long time. The eyes were reduced by weathering and age to indistinct dimples, barely recognizable among the coarse, pebbly aggregate. The crown of thorns was worn down to the nubs, and if you didn't know the story, you might not even recognize the nubs as thorns. Finally, dejected, I turned to walk away, quietly saying to myself that it had been just too long to recognize this statue's true form; virtually every important feature had been lost over the countless years since its creation.

As I turned to walk away, I felt a hand on my shoulder! Shock waves of electricity shot up and down my spine, and my body became one single mass of goose flesh. I turned to face the statue and saw a hand attached to the statue's now-outstretched arm. I knelt before the statue, taking the statue's hand in mine. I spoke to the stature in a halting, fearful tone, "What would you have me do?" The statue did not answer but gripped my hand even harder. I could feel the warmth of blood circulating through the statue's hand. I instantly woke up, shaking uncontrollably and covered in sweat.

Today, most people on our earth look to either religion or to science as the ultimate source of truth in their lives. Jesus taught that a house divided cannot stand, and so it is with truth. If our prime source of truth is religion and science contradicts this truth, who is right? If our prime source of truth is science and religion contradicts this truth, who is right? Many people in today's world seem to have an ability to believe both religious and scientific truth at the same time and simply ignore the contradictions between them. While this may be a very practical solution to the problem for these people, it is intellectually and spiritually unsettling for many others, including the authors. Let us consider ways of resolving this dilemma.

Science and religion may seem to have very little in common, but in fact, they share a common ancestor in the tribal shaman of our forebears. The shaman's function was (and is) to help the tribe to function by tapping into the unseen world of spirit. There he would find answers to questions concerning hunting and planting, when to make war, and when to make peace. The shaman was also the tribe's healer. As cultures evolved from their tribal beginnings, the shaman's role became divided into two functions that, for lack of better words, can be called workers of magic and spiritual traditionalists. Science has evolved out of the magic-working tradition, and religion has evolved along a parallel path of the spiritual traditionalists. Let's consider the characteristics of each.

During the Middle Ages, workers of magic in Europe became grouped into those who practiced alchemy, those who practiced astrology, those who practiced other forms of divination, and those who practiced healing. Magic working was a solitary pursuit. Quite often a master worked alone with perhaps one or two apprentices. The results of magic working were often shared with people throughout a community but always on a one-to-one basis. This kind of work was never meant to be made very public because the work itself often encountered significant prejudice from the community's authorities.

The goal of a magic worker was to understand and control forces in the natural world. The alchemist worked to change a substance of lesser value into a substance of greater value. The astrologer watched the movements of the stars and planets to gain an understanding of what future events might befall individuals based on their birth dates. A diviner would perform seemingly random events (such as drawing cards or rolling dice or coins) to answer a particular question or predict future events. A healer might use herbs or other substances combined with ritual prayers to bring about a healing.

However, with the coming of the Renaissance (toward the end of the fifteenth century) magic workers started to develop a common language that evolved into what we now call mathematics. With the introduction of mathematics, the magic tradition itself began to evolve into what today we call science. Using their newfound mathematical tools, these new magicians/scientists began to explore the natural world using the time-tested scientific formula of measurement, hypothesis, confirmation, and replication, all leading up to the development of a theory of the natural world. During this period of change, alchemy evolved into physics and chemistry, astrology evolved into astronomy and cosmology, divination evolved into the mathematics of predicting future events, and healing evolved into medicine. Today it is a little-known fact that many of the famous scientists from the past were also workers of magic. For example, Isaac Newton[3] spent more time on alchemy than he ever spent on mechanics and calculus. Johannes Kepler[4] was an astrologer.

The goals of these new scientists remained essentially the same as the goals of the magic workers: achieving the understanding and control of the

---

[3] Isaac Newton (1642–1727) was an English physicist born in Lincolnshire. Newton revolutionized the science of his day by developing a mechanics that explained the orbits of all of the solar system's planets. He also developed the mathematics of calculus to provide a mathematical underpinning for his mechanics. Newton also spent considerable time at the magical practice of alchemy. Later in life Newton served as warden of the royal mint.

[4] Johannes Kepler (1571–1630) was a German astronomer (and astrologer) who produced very accurate planetary charts that led him to developing a system of orbital mechanics that now bears his name.

forces of the natural world. In our day, the pursuit of these goals is carried out jointly by science and technology. Nevertheless, the roots of science remain in the magic of the past, and the goals of today's science remain essentially unchanged from the goals of the ancient magic workers. Arthur C. Clarke,[5] the father of the communications satellite, once remarked that the technology of an advanced civilization would always appear as magic to a less-advanced civilization. The difference between science and magic may simply be in the eye and understanding of the beholder.

Religion, on the other hand, has always been a very public endeavor. Religion teaches people to trust in God and to know they are a part of something larger than themselves. Religion teaches that we human beings are at the very center of God's creation. For the religious person, the point of God's creation is to provide a home for God's greatest creation: namely us.

Religion has always been tied closely to cultural identity. All cultures use religious rituals and ceremonies to instill in their members a sense of transcendence that teaches the ethics and values of the culture in a way beyond words. The leaders of a particular culture often portray themselves as God's appointed leaders whose role it is to govern God's people on earth. It is religion that gives people a sense of purpose and belonging that serves to bind their culture together. If faith in a culture's religion begins to fade, the culture as a whole suffers pain, a sense of dislocation, and a lack of harmony. To the religious person, human beings are the centerpieces of God's creation, and their culture is the fulfillment of God's plan on earth.

In the Western culture of today, there exist many seeming conflicts between the religious and the scientific points of view, bringing with them a loss of self-evident, reliable authority. For instance, as opposed to religion, science teaches that humans do not occupy any special place within the natural world. In fact, to many scientists, our very creation may just have been an extremely fortunate accident. To the religious person, such talk is blasphemy. Scientific truth teaches us that our earth is not at the center of creation but is located in a relatively obscure corner of the cosmos. Science teaches us that human beings have evolved naturally from lower forms of life without the help of any kind of divine intervention.

---

[5]   Arthur C. Clarke (1917–2008) was an English communications engineer and science fiction writer. He is the inventor of the communications satellite.

Science teaches us that our earth, our solar system, and our universe are far older than anything we can infer from the biblical accounts of creation.

However, science teaches us absolutely nothing about ethics and values. Science also teaches us nothing about our culture's special role in God's creation, except perhaps to recognize that we as individuals have a special proclivity to spread our own genetic material as far and as wide as possible. However, there still remains the problem of truth! Is there a single, whole, and unified truth, or are there two truths—the truth of religion and the truth of science? We remember the words of Jesus: "A house divided against itself cannot stand" (Matt. 10:25). What will become of a divided truth? Could a divided truth be a part of God's plan?

Today many, if not most, scientists are either deists or atheists. Deists believe that God created the universe somewhere in the dim, distant past and then stood back from creation and let the universe run by itself, as if it were a kind of clockwork, evolving according to the deterministic laws of physics, chemistry, and biology. On the other hand, most religious people are theists. Theists believe that God cares about creation on a very personal level. The God of the theist is with all of us on a day-to-day basis, sharing our victories and comforting us in our defeats. The God of the theist has bestowed the gift of free will on all humans, charging us to do what is good and to renounce what is evil. To the theist, God is with us all on a day-by-day, minute-by-minute basis. Yet God is in no way a part of our material universe. Paradoxically, God stands apart from the created world but all the same is always there for us.

We the authors think it is very important, in the interest of personal wholeness, for scientific and religious people to make sacrifices at each other's altars. Like the Trinitarian God of Christianity, truth may have multiple faces, but it can never be divided. Each of us must come to know and respect the other's point of view for the wholeness of truth to contain all.

Our purpose in writing this book is to explore ways for achieving this sense of wholeness by searching for the common ground upon which religion and science are both based. To this end, we would like to propose a very important question: how does the God of the theists communicate a plan for humanity to each of us? A constantly ongoing divine communication channel between God and humanity is absolutely essential from the point of view of the theist. But to the scientist, it is extremely important that God's communication with humanity be accomplished without breaking any of the scientific laws that form the basis for God's creation of our natural world. It is these very laws that are at the heart of science. They reside in the hearts and minds of all scientists.

If God were to break God's own laws for the purpose of communication, the whole of science would fall because the laws of the natural world would no longer be universal and dependable. The laws of science could no longer be considered universal but would be demoted to the level of temporary and changeable. The question of divine communication becomes absolutely fundamental to both the religious person and to the scientist who believes there is a theistic God. As we explore this question over the course of this book, it will become clear why each side must be willing to sacrifice at the altars of the others if a consensus of universal, undivided truth is to be achieved.

Many of us have the gift to see forms in clouds on a summer day. Fran, one of the authors of this work, recently witnessed a beautiful cloud formation that appeared to be a dancing Albert Einstein whose pounding boot heels seemed to be beating out a flamenco rhythm upon a wooden stage of clouds as his right hand stretched ever higher, pointing upward toward the zenith. However, as is the way of all cloud formations, in a split second the clouds had changed into a soaring eagle, taking wing up and up into the afternoon sky. Once the dancing Einstein became the eagle, it became extremely difficult to recapture the perception of the dancing Einstein. But this is exactly the kind of challenge we encounter when we attempt to simultaneously perceive the unity of the truth of science and the truth of religion. We are forced to stretch ourselves to seeing the dancing Einstein and the soaring eagle simultaneously! The rest of this book will be a quest for accomplishment of this feat!

CHAPTER 1

THE BIG BANG THEORY, WHERE THE TRUTHS OF SCIENCE AND RELIGION HAVE ALREADY MET

How the Big Bang Theory Came About

We start our quest at the beginning, at a point where the truth of science and the truth of religion have already overlapped—perhaps by accident, perhaps by intent—in a way that is quite mysterious and largely unknown to most scientists and religious people alike. The story starts out with the equations of general relativity that were developed by Albert Einstein[6] in 1915 and with a suggestion from a Belgian Jesuit priest and physicist, Father Georges Lemaitre,[7] of a novel solution to these equations. Father Lemaitre was the first (in 1927) to call to the attention of the world's

---

[6]   Albert Einstein (1879–1955) was a German (and Swiss and American) physicist. Einstein was perhaps the most famous scientist of all time. He singlehandedly changed science forever with his theories of special and general relativity. Einstein's work forms the core of our human understanding of how the entire cosmos functions. Einstein received the Nobel Prize in 1921 for his explanation of photoelectric effect, which constituted the basis of the science of quantum mechanics.

[7]   Georges Lemaitre (1894–1966) was a Belgian physicist and Jesuit priest. It was Father Lemaitre's solutions to Einstein's equations of general relativity that led to the eventual discovery of the big bang theory of the universe's beginning.

1

physics and cosmology communities that a theory of cosmology could be constructed by identifying a point in time when the universe began. A little later, Alexander Friedman[8] in Russia came to a similar conclusion.

The steady state theory of the universe that was most favored by the scientists of the time was unacceptable to theologians because it elevated the material universe to the position of being coeternal with God. However, Father Lemaitre found a solution within Einstein's equations of general relativity that predicted the existence of a definite point in time when our universe began. If this were true, our universe was not eternal, answering the objections of theologians. Once our universe came into being, it embarked upon a period of expansion that continues to the present day. This picture of our universe's beginning is remarkably similar to the creation story in the Bible (found in Genesis 1). Father Lemaitre never publically stated his intention to build a bridge between the biblical account of creation and the scientific account of creation, but it is hard to imagine that a man of the cloth would not have this motive in mind.

Initially, the physics-cosmology community did not take Father Lemaitre's proposal very seriously. But gradually, a fantastic story of discovery began to unfold as evidence mounted that favored Father Lemaitre's ideas. First there was the Hubble red shift,[9] and later, to clinch the argument, there came the discovery by two Bell Laboratories communication engineers (in 1962) of cosmic microwave background noise.[10] Interestingly, the term "big bang" was initially used as a term of derision by Fred Hoyle, a British steady state cosmologist. Like many other names of derision, this one has stuck. We will now tell the whole story of the discovery of the big bang theory of cosmology in the detail it deserves.

---

[8]   Alexander Friedman (1888–1925) was a Russian physicist who, like Father Lemaitre, solved Einstein's general relativity equations in such as way that a growing universe with a definite beginning was a possibility.

[9]   Edwin Hubble (1889–1953) was an American astronomer who was the first scientist to observe that a galaxy's speed as it recedes from earth is proportional to its distance from earth. This phenomenon is known as the Hubble red shift.

[10]   The cosmic microwave background is all that remains today of the cosmic fireball that occurred at the universe's birth during the big bang.

Evidence in Favor of the Big Bang Theory of the Universe

Perhaps the most significant event in human history happened approximately 13.5 billion years ago. It is called the big bang. Cosmologists now understand the big bang as a point along the timeline of history when time itself began. No time existed before the big bang; in fact, nothing (i.e., no things) existed prior to the big bang. No space, no time, and perhaps no laws of physics existed; nothing of a material nature existed in any sense of the word.

Physicists call the moment when the big bang occurred a singularity.[11] A singularity is a lot like a black hole, except it works roughly in reverse. A black hole swallows up all the surrounding matter and energy (within its local event horizon) by the immense gravitational attraction of its point-like grasp. On the other hand, the big bang's[12] singularity started out as a point and grew outward, serving as the genesis for all things within our universe. All of the matter and energy, all of the laws of physics and chemistry, and even space and time themselves expanded outward from that first infinitesimal moment at the point of creation called the singularity. Before the singularity, there was no matter, no energy, no space, and no time; but after the singularity, the potential for all things came into being within that very instant of the singularity.

How do we know the big bang singularity theory is true? There are several competing cosmological theories of creation, not to mention all of the creation stories contained in the world's religious traditions. What is so special about the big bang theory that drives us to take it so seriously? The answer is quite simple. There exists experimental evidence (and lots of it), of the kind that scientists love so dearly, in support of the big bang theory. What is so special about the big bang theory is that religious people of all cultures are able to recognize their own creation stories within this scientific account of creation.

---

[11]   A singularity was the infinitesimal universe at the moment of the big bang. In the singularity, all of the future universe's matter and energy was concentrated at a single point of zero size.

[12]   The big bang is the cosmological theory that holds that our universe had a definite beginning approximately 13.5 billion years ago.

Before discussing the evidence in support of the big bang theory, let us first discuss the concept of evidence in a general way. Science, as we know it today, got its start in the seventeenth century CE. (Of course there was a wealth of scientific literature produced by the ancient Greeks and Romans, but it wasn't until the Renaissance in Western Europe that science adopted the universal language of mathematics.) The cornerstone of science is in the gathering of experimental data about the natural world and applying reason and logic to this evidence for the purpose of developing new theories explaining the observed natural phenomena. Prior to the seventeenth century, what we in the twenty-first century think of as the disciplines of natural science, theology, and philosophy did not exist as separate branches of knowledge anywhere on earth. Prior to the scientific revolution of the seventeenth century, if a new finding in the natural sciences could not be understood by reason and logic, scientists would defer to theology by assuming they were dealing with an act of God.

However, by the seventeenth century, scientific pioneers like Copernicus,[13] Bruno,[14] and Galileo[15] were coming to understand that this method was a copout hindering the growth and advancement of the natural sciences. It was all too easy for the scientists of that day to refuse to dig further into the problem of finding a rational/logical explanation when theology held out such an easy explanation. Some mighty battles were fought between budding scientists and the church over such issues as the Copernican sun-centered solar system, and of course in the end, the church ultimately lost these battles.

---

[13] Nicolas Copernicus (1473–1543) was the Polish astronomer who is responsible for the theory of a sun-centered solar system in a historic period when most educated people believed in an earth-centered solar system.

[14] Giordano Bruno (1548–1600) was an Italian scientist whose views relative to the sun-centered solar system led the church to regard him as a heretic. Bruno would not change his views and was burned at the stake as a heretic by a church court.

[15] Galileo Galilei (1564–1642) was an Italian physicist and the inventor of the telescope. Galileo was a strong supporter of the sun-centered solar system and was investigated by the church on charges of heresy. Galileo recanted his views to save his life. He spent the rest of his life under house arrest.

Today an understanding of the natural world is considered by most educated people worldwide to be the sole domain of science. When those who are trying to bring God back into the picture dare to intrude on the sacred domain of science, there is controversy. Is this progress? Maybe yes, maybe no. Perhaps in the twenty-first century science has taken for itself the same power and influence that was held by the church in the seventeenth century.

Hubble's Red Shift

Returning to the big bang theory, let us examine the experimental evidence in its favor. In the 1920s and the 1930s, an astronomer named Edwin Hubble made some very significant discoveries while observing with the new telescopes that had recently been constructed in the mountains east of Los Angeles, California. Hubble used an optical device called a diffraction grating spectrometer to accurately measure the exact frequency components associated with starlight he was observing. Hubble made the daring assumption that all atoms and molecules anywhere in the universe that were of identical composition to those same atoms and molecules on earth behave in exactly the same way as their earthbound cousins (i.e., hydrogen is hydrogen anywhere in the universe). Hubble also developed a technique for judging a galaxy's distance from earth based on its observed brightness.

By making use of the above assumption that distant galaxies are composed of the same type of atoms we experience here on earth, Hubble noticed that certain spectral emission lines of many well-known elements, such as hydrogen, were always shifted to *lower frequencies* when they were observed originating from distant galaxies. Since Hubble's observations always indicated a frequency *reduction*, he called this phenomenon a red shift[16] because the color red is associated with the lowest frequency component of visible light. Next Hubble did something very creative: he used a piece of graph paper to plot each galaxy's red shift versus its distance from earth (as determined from the brightness criteria). The

---

[16]   Red shift is a phenomenon first observed by Edwin Hubble in which any galaxy is receding from earth with a speed that is proportional to its distance from earth.

results were astounding. All of Hubble's data points lay along a tightly grouped straight line. This discovery told Hubble that a galaxy's red shift was directly proportional to its distance from earth, with the most distant galaxies having the most red shift.

Next Hubble made an association between the red shift and the now-familiar Doppler frequency shift[17] that is associated with all types of wave phenomena emanating from moving objects (i.e., cars, trains, or galaxies). In all Doppler shift phenomena, observers on objects moving apart from one another experience a decrease in wave frequency. A familiar example of a Doppler red shift is a stationary observer listening to the whistle of a train moving away from a grade crossing.

However, when objects move toward each another, the observers experience an increase in frequency. Hubble never observed an increasing light frequency from any of the distant galaxies he observed, which is a very significant result because it indicates that all galaxies are moving away from Earth. The amount of Doppler frequency shift depends on the relative velocities of the two objects; the greater the relative velocity, the greater the Doppler frequency shift. Therefore, the more-distant galaxies having the greater red shift are moving away from earth more rapidly than the closer galaxies with less red shift.

Later cosmologists came to recognize that a simple Doppler shift explanation of Hubble's red shift data was not exactly correct. In fact, the galactic red shift is really caused by the expansion of space itself that took place after the big bang. Because of the expansion of space during the early universe, all waves, like light waves, are stretched in wavelength, lowering their frequency (i.e., they are red shifted). The conclusion science now draws from Hubble's red shift is that all galaxies are moving away from earth because all of the frequency shifts observed by Hubble were decreasing (red shifting). Since the amount of red shift is directly proportional to the galaxy's distance from Earth, the most-distant galaxies

---

[17] Doppler shift is a phenomenon in which a moving object's wave emissions are seen to change frequency by a stationary observer. The amount of frequency shift is directly proportional to the relative velocity of the moving object. An approaching object causes an upward frequency shift, and a receding object causes a downward frequency shift. Hubble's red shift can be thought of as Doppler-shifted light coming from distant galaxies.

must be moving away from Earth most quickly and the closest galaxies are moving away from Earth more slowly.

Hubble's red shift phenomenon suggests a universe that is behaving like raisin bread cooking in an oven. As the bread rises, all of raisins move away from each other, with those raisins that are farther away from each other moving more quickly than those raisins that are close together. Since the Hubble universe is constantly expanding, it strongly suggests the existence of a point in the past when the universe was very small, compact, perhaps a point like singularity (i.e., the big bang!).

At the time Hubble was making these measurements, most scientists were still in favor of the steady state theory of the universe, which is a universe that neither grows nor decays. Hubble's data was enough to convince Einstein of the validity of an expanding universe. Einstein went on to modify his own theoretical understanding of the universe to include Hubble's observations of an expanding universe, as had been proposed earlier by Father Lemaitre. Slowly, over time, the scientific community became convinced that an expanding universe was not only possible but was the explanation that offered the best fit for the experimental data. Whether science was willing to admit it or not, the cosmology theory that best fit the experimental data was the theory in closest agreement with the creation stories of the Bible!

Thermodynamics, Entropy, and the Big Bang

The second piece of evidence in support of the reality of the big bang has nothing to do with either astronomy or cosmology. In fact, this evidence is from the far-flung science of thermodynamics.[18] Thermodynamics is the universal study of energy interactions within materials in the presence of heat. The first law of thermodynamics is also known as the law of conservation of energy.[19] This first law tells us that the overall sum of all of the energies contained in the universe is a constant, never increasing or

---

[18] Thermodynamics is the science of heat energy transfer within physical systems.

[19] Conservation of energy is the first law of thermodynamics, which holds that in any physical interaction energy can be neither created nor destroyed but only transformed. At the cosmic level, this law means that the total energy of the universe is fixed and can never change.

7

decreasing but remaining absolutely constant for all time. However, energy is constantly being transformed as it is traded back and forth among all the various components of the universe. Nevertheless, the total energy (i.e., the sum of all the energy in the universe) must remain a constant that is exactly equal to the energy contained in the big bang singularity at the moment of creation.

However, it is the second law of thermodynamics that is really interesting from the point of view of providing evidence in favor of the big bang theory. The second law of thermodynamics is also called the law of increasing entropy.[20] Entropy, which is related to chaos and disorder, is a physical property of every physical system that can be difficult to understand. What is difficult to grasp about entropy is that it seems so far removed from our everyday experience, but it really isn't. Calculating the entropy increase of a closed thermodynamic system just requires dividing the energy input being delivered to system by the system's temperature. It is easy to show mathematically that in every case of a thermodynamic energy exchange, the entropy of a closed system always increases. This law is equally true for steam engines, transistors, dogs, moon rockets, human beings, and galaxies.

The law of increasing entropy is an absolutely universal thermodynamic property of each and every manifestation of matter and energy in our universe. Since every conceivable type of physical interaction leads to an increase in entropy, the overall entropy of the universe must always be increasing! This means the total entropy of the universe has been steadily increasing since time began. If we could calculate the rate at which the universe's entropy is increasing, we could, in principle, project this rate of entropy increase backward to a point in time when the total entropy of the universe was exactly zero.

The implication of this thought experiment is that no physical (thermodynamic) activity could have existed before some point in time in the distant past, implying that at some point in time the entropy of the universe was exactly zero, which would have been the moment of the universe's creation. No activity of any kind could have taken place before this point in time. It is not unreasonable to associate this zero entropy point in time with what cosmologists call the moment of the big bang's singularity.

---

[20] Entropy is a measure of physical disorder. The second law of thermodynamics holds that the total entropy of the universe is always increasing.

Before this point in time (t=0) there could be no thermodynamic activity of any kind, implying there was no universe before t=0. As we shall see in later chapters, the second law of thermodynamics is one of most solid pillars in the bedrock of science. It is the big bang theory of creation alone—among all of the other cosmological theories—that is completely consistent with the implications of the second law of thermodynamics!

The Cosmic Microwave Background Noise

A third, and perhaps the most compelling, piece of evidence for the reality of the big bang is the cosmic microwave background noise. The cosmic microwave background noise was discovered in 1962 by two Bell Laboratories communications engineers, Robert Wilson and Arno Penzias.[21]

Wilson and Penzias were working together on the development of a communications satellite earth station intended to listen for radio signals from the newly deployed communications satellites. The pair of engineers found that when they turned their massive horn antenna toward a quiet part of the sky (away from any known galactic noise sources), there was always a small but discernible excess noise temperature of about 2.7 degrees Kelvin. Noise temperature is the way communications engineers and radio astronomers describe an earth station's sensitivity in receiving weak signals from outer space. It did not matter where in the sky their antenna was pointed; the result was always about 2.7 degrees Kelvin. This finding dismayed the pair of engineers so much that they began to take extreme measures to explain the discrepancy. They even went so far as personally scraping the pigeon droppings off of the inside of the horn antenna. But nothing worked; no matter what they did, there remained an unexplained 2.7 degrees Kelvin excess noise temperature with the system.

Lucky accidents often precede great scientific discoveries, and this story is a prime example of such a lucky accident. One of the Bell

---

[21] Robert Wilson (born 1936) and Arno Penzias (born 1933) are American communications engineers and codiscoverers of the cosmic microwave background noise, which provided the strongest evidence of the reality of the big bang theory of the universe's beginning.

communications engineers had a friend, Robert Dicke,[22] who was a cosmologist at Princeton University. By chance, the friend just happened to be working on a new cosmology theory called, of all things, the big bang theory. A meeting was arranged at Princeton between the communications engineers and the Princeton cosmologists. To everyone's amazement and joy, the Bell Laboratory team's measured 2.7 degrees Kelvin of excess noise temperature was exactly what the Princeton cosmologists were calculating as the temperature that would remain today as the result of the big bang's primordial fireball (which occurred for a brief moment after the big bang's singularity). Of course, it took 13.5 billion years of universe expansion and cooling to reduce the fireball's temperature (by a gradual process of red shifting the cosmic fireball's visible light down in frequency to the microwave radio frequencies used by the Bell Labs team) to exactly 2.7 degrees Kelvin!

This incredible agreement seemed to be more than just lucky chance. For the first time, there was really hard, direct evidence of the reality of the big bang! There was much heated debate over the meaning of these measurements, but in time, most cosmologists came to accept the validity of the cosmic microwave background noise as irrefutable, direct evidence, proving beyond any reasonable doubt the validity of the big bang theory.

For their efforts in this discovery, the two communications engineers went on to share a Nobel Prize in physics. Since then, even the pope has spoken approvingly of the big bang theory because it sounds so unmistakably similar to the biblical account of creation. The big bang theory, which was first proposed by a priest, had now arrived at universal acceptance within the world of science. A rare and unique occurrence had taken place; religion (for a change) had lit the way forward for scientific progress. Scientific truth and religious truth had—for just this moment—come into perfect alignment and were, within the narrow confines of the big bang, truly united. If this situation can happen once, it can happen again.

---

22   Robert Dicke, an American physicist and cosmologist, is best known for his work on quantum mechanics, interference effects, lasers, and big bang cosmology.

Grasping for the Big Picture

In the final analysis, the understanding that has emerged from this work is that our universe was created about 13.5 billion years ago and has been expanding ever since. At present, the most-distant objects from Earth are greater than 13.5 billion light years away. Most of the mass and energy present in the initial big bang singularity has over the eons condensed into stars, planets, and galaxies. All of the elements in the periodic chemical chart have been forged within the nuclear caldrons of stars and supernovas. The number of stars and galaxies in the universe is of course finite (because the entire universe is finite), but there are such a large number of stars and galaxies that their number defies human comprehension. One estimate is that more than 100 billion galaxies exist within our universe.

Our place within this cosmic vastness seems very small and remote, making it hard for us to envision, as the Bible teaches us, that all of this has happened for our benefit alone. Could it be that other life and lives are taking place elsewhere in the universe? Perhaps we are not alone in God's creation. This very question has tantalized humans ever since the beginning of the human race, and there is still no real answer to it from either science or religion. We find ourselves feeling unsure of whether it seems more reasonable to think of God creating such a vast universe for our benefit alone, or if it is more reasonable for us to expect to have counterparts spread across the entire expanse of the universe's vastness.

The scientific discovery of the big bang theory seems remarkably consistent with the creation story in Genesis 1. However, our aloneness within the vastness of creation is really hard to accept. Surely God would have put this vastness to more useful purposes than simply providing a home for our human race that resides in a tiny corner of creation. There is also a matter of timing and content when comparing the big bang to Genesis 1. The creation story of the big bang occurred over a period of 13.5 billion years; however, the Genesis 1 creation story happened within the period of a few days at the most. Our scientific understanding of the universe tells us that billions of galaxies exist, each of a size that is measured in hundreds of millions of light years (a light year is the distance that light can travel in one earth year), and each containing countless billions of stars. These astronomically great numbers and sizes are hard for us to associate with the description of the stars as the lights in the night sky in the Genesis 1 account.

However, we can safely say that the scientific account of the big bang is similar to the Genesis 1 story of creation as long as we read the Genesis 1 account as a mythic poem of creation rather than as a literal history of our universe. Again we see how the creation stories of religion and science have a problem communicating with each other because they speak different languages. Religion speaks the language of myth, poetry, ritual, music, and metaphor while science speaks the language of observation, measurement, and mathematics. We must always bear in mind that science and religion are speaking in their own languages about the same event. Religion speaks the language of the heart, which is emotional, while science speaks language of rational thought, which is factual.

Each of us needs to internalize both descriptions of creation to have a fully informed understanding of the event. Remember how in the preface of this book, the clouds changed from a dancing Einstein into a soaring eagle. To grasp the whole of the truth, it is necessary to grasp both the religious truth and the scientific truth simultaneously. The big bang theory demonstrates how science and religion can exist simultaneously as facets of the same truth. This way of experiencing the unity of truth can happen again and again. Now let us see where, and how far, we can take this concept. In the next chapter, we will tackle the greatest of all scientific mysteries—the meaning of quantum mechanics.

CHAPTER 2

THE QUANTUM REVOLUTION

Determinism Is Replaced by Quantum Chance

No area of science has been more plagued by doubts, confusion, and mystery than the branch of physics called quantum mechanics. Niels Bohr,[23] one of the principal architects of quantum mechanics, once remarked that anyone who has not been profoundly disturbed by quantum mechanics does not really understand what quantum mechanics is saying. Nobel Prize winner Richard Feynman[24] once said that you never really understand quantum mechanics; you just come to accept it.

While many practicing scientists take an attitude of "shut up and calculate," the more philosophically inclined scientists have long struggled to truly understand what quantum mechanics is really telling us about how our universe works. But along the lines of making a silk purse out of a sow's ear, we will endeavor to find within the mysteries of quantum mechanics the many jigsaw puzzle pieces that can serve as ingredients for unifying scientific and religious truth. In fact, by looking at quantum mechanics from just the right angle, it might be possible to solve some really nasty problems in both science and religion at the same time! So

---

[23]   Niels Bohr (1885–1962) was a Danish physicist who developed a theory of the atom that bears his name and led a movement that has come to be known as the Copenhagen Interpretation of quantum mechanics. Albert Einstein was deeply troubled by Bohr's formulation of quantum mechanics. Bohr received the Nobel Prize in 1922.

[24]   Richard Feynman (1885–1962) is an American physicist best known for his work on quantum electrodynamics. Feynman received the Nobel Prize in 1965.

sit back, relax, and enjoy the ride. We assure you that the experience will be worth it as we go about collecting gleaming nuggets of wisdom while panning these quantum waters.

The quantum revolution in physics is now just over one hundred years old. This truly revolutionary development in physics singlehandedly overturned the neat deterministic clockwork universe of Newton and Laplace.[25] Quantum mechanics is arguably the leading intellectual achievement of the twentieth century. The major outcome of this revolution has been to change the scientific predictive certainties of determinism into a long string of statistical "maybes" and "it all depends."

Max Planck[26] and Albert Einstein jointly founded quantum mechanics in the years immediately following the turn of the twentieth century. What was to become quantum mechanics was preceded by the recognition that in the world of the very small (the world of atoms and molecules), energy is transferred only in discrete steps called quanta. In one of Einstein's world-changing 1905 scientific papers (explaining the photoelectric effect), Einstein recognized that if light is to carry energy in tiny discrete quanta (later called photons), then light must have a particle nature (as Newton had earlier insisted) as well as a wave-like nature. James Clerk Maxwell[27] proved the wave-like nature of light to be true about thirty years earlier.

At the same time Einstein was working out his investigation into the nature of light, a young graduate student in France named Louis de Broglie was working on his own hypothesis. He concluded that if Einstein's light quanta could act both as waves and as particles, then particles such as electrons and protons—which are the building blocks of atoms and molecules (and of all matter)—also have a dual particle-wave nature. De Broglie's concept of "matter waves" was soon observed in the laboratory, bolstering the believability of Einstein's paradoxical particle-wave hypothesis.

---

[25] Pierre-Simon Laplace (1749–1827) was a French mathematician and astronomer. He was best known for his work on probability theory.

[26] Max Planck (1858–1947) was a German physicist. Together with Einstein, Planck was one of the founding fathers of the new science of quantum mechanics.

[27] James Clerk Maxwell (1831–1879) was a Scottish physicist best known for his groundbreaking work in classical electro-dynamics and statistical mechanics. The four equations that bear his name define classical electro-dynamics.

Soon de Broglie and Max Born[28] in Germany were asking questions like, "What is really waving within these matter waves?" It occurred to de Broglie and Born, who were working separately, that perhaps it was the particle's very existence that was waving. They finally concluded that the wave nature of subatomic matter required the existence of a new kind of wave amplitude that was capable of representing the chance (or statistical probability) of a particle being found at a given location at some particular point in time.

Based on de Broglie's and Born's ideas, the element of chance entered the picture as a factor in the equations describing the behavior of subatomic matter and energy. In fact, chance turned out to be at the very core of what science could even hope to understand about matter and energy interacting in the subatomic world. From this point forward, science had to grudgingly admit to a fundamental inability to simultaneously measure, with absolute certainty, a particle's position and its momentum. (They also had to admit to similar inabilities to simultaneously measure other pairs of dynamic variables, such as time and energy.) Calculating certain "probabilities" that particular values of these properties might occur was all that could ever be hoped for from the new field of quantum mechanics. It seemed to the scientists who were making these discoveries that the very existence of the particles was the thing that was waving. Measuring quantum mechanical wave properties of subatomic matter was in no way similar to any kind of "normal" wave phenomena measurements that could be verified in the laboratory by customary scientific measurements.

But What Does It All Mean?

As time went by, more scientists got into the act, rapidly moving the developing field of quantum mechanics forward along the lines of several competing schools of quantum thought. Perhaps the most advanced and creative of these schools of quantum mechanical thought was located in

---

[28] Max Born (1882–1970) was a German physicist who was one of the leading lights in the development of quantum mechanics. He is best known (together with Louis de Broglie) for introducing the concept of probability waves into quantum mechanics.

Copenhagen, Denmark, and functioned under the leadership of Niels Bohr.

Bohr had gathered around himself a small but brilliant staff of young scientists, who Bohr personally directed and mentored. These brilliant young scientists were mostly young and in their early twenties. Their work was often called Knabenphysik, which is German for "boy-physics." They led the charge into the formal development of the rapidly maturing field of quantum mechanics. Soon to become famous were young scientists like Werner Heisenberg and Wolfgang Pauli,[29] who joined Neils Bohr in Copenhagen for what many felt would be the final assault on the formal development of quantum mechanics, which took place in the early 1930s.

It was during this period that Werner Heisenberg proposed his now-famous uncertainty principle. It was Heisenberg's uncertainty that mathematically formalized the degree to which observers of the subatomic world are absolutely and fundamentally denied perfect knowledge of the properties of matter and energy. Another famous principle that was developed in Copenhagen during this period was Wolfgang Pauli's exclusion principle. It formalized nature's restrictions on the duplication of a particle's wave state by a second particle. Pauli's exclusion principle gave science new clarification about the rules that govern how electrons fill in the atomic energy levels associated with atoms of all kinds. Pauli's exclusion principle places the periodic chart of chemical elements on a firm theoretical basis.

During this period (late 1920s to the early 1930s), Bohr himself proposed a concept he came to call complementarity. It explained how and why the measurement of subatomic particles is restricted to measuring some of the particle's properties some of the time but not all of a particle's properties at any time. Bohr came to understand that it is the configuration of the laboratory measurement equipment that defines which properties of a quantum mechanical system can or cannot be measured. Bohr felt strongly that the measurement restrictions on what can and cannot be measured that in a very fundamental way define the reality (or unreality) of a particle's properties.

---

[29] Wolfgang Pauli (1900–1958) was a German physicist and the discoverer of the (Pauli) exclusion principle, which is an extension of quantum mechanics that explains the periodic chart of chemical elements.

In Bohr's view, properties that cannot be measured are for all practical purposes nonexistent properties, at least in the context of a particular measurement system's configuration. Therefore, how one configures an experiment has a major impact on determining what properties of a quantum entity do and do not exist in the subatomic world. As Bohr put it, quantum mechanical states contain the potential for all properties to exist at once, but certain properties can actually be observed only in special measurement configurations. And never can all properties be observed at the same time.

In his doctrine of complementarity, Bohr came to view the state of a quantum mechanical system as the superposition of many potentially overlapping properties that only become real when they are observed in a specially configured laboratory experiment (that is designed to observe that particular property). Therefore, according to Bohr, reality in the subatomic world is jointly determined by a superposition of waves that make up the wave state of the particle and the specific configuration of the experimental system making the measurement. Wave state is a state of nature, but measurement is interference by the human will.

The Bohr-Einstein Debates

The theoretical directions taken by Bohr and his Copenhagen school of quantum mechanics greatly frustrated Albert Einstein. Although Bohr and Einstein had been and remained close, lifelong personal friends, they became intellectual sparring partners at most of the physics conferences that were held across central Europe throughout the late 1920s and early 1930s.

Not long after this time period, Einstein, who was Jewish, emigrated to the United States to escape the rise of Nazism in Germany. The crux of Bohr-Einstein debate was that Einstein, a strict determinist, could not accept a quantum mechanical world (according to Bohr) where it was fundamentally impossible to gain complete scientific knowledge through the measurement of a subatomic particle's properties. Concerning the fundamentally probabilistic nature of Bohr's quantum mechanics, Einstein famously said, "God does not play dice," to which Bohr retorted, "Albert, who are you to tell God his business?"[30]

Over the next five years, Einstein searched frantically, but in vain, for what he called "hidden variables." He hoped they would be capable of recovering the lost information that was being denied to the observer by Bohr's version of quantum mechanics. Einstein struggled for the rest of his life to try to disprove the Copenhagen school's interpretation of quantum mechanics but with little success. During his valiant but hopeless search for logical contradictions in Bohr's theories, Einstein spoke a lot of words and spilled a lot of ink, but to no avail.

Einstein truly played Don Quixote as he jousted at Bohr's Copenhagen windmills. The hidden variables that Einstein so hoped for were never to be found. Einstein was never reconciled to Bohr's theories, nor was he ever able to disprove them, a fact that brought much sadness to both of these great scientists. However, it is very important to understand and remember that by playing the role of gadfly, Einstein pushed Bohr to the limit to defend his vision of quantum mechanics. Quantum mechanics and all of science is the better for it! Most scientists of the day (and also today) sided with Bohr in the ongoing debate. Paul Ehrenfest,[31] a close personal friend of Einstein's, once said privately, "Albert, you are beginning to sound more and more like the critics of your own relativity theories from so many years ago."[32] To no avail, Einstein would never give in and accept Bohr's theories, and he did not ever succeed in casting reasonable doubt on their validity. Till his death in 1955, Einstein remained gravely disturbed by the meaning and direction taken by the Copenhagen school of quantum mechanics.

Einstein was not alone in his concerns. Other competing interpretations of quantum mechanics have continued to pop up since the wellspring years of the 1920s and '30s. Chief among these are the many worlds interpretation

---

[30]   Letter from Einstein concerning his conversations with Neils Bohr to Max Born, December 4, 1926, Einstein Archive 8–18.

[31]   Paul Ehrenfest (1880–1933) was a German physicist and close friend of Einstein. Ehrenfest is best known for his work in statistical mechanics.

[32]   Walter Isaacson, *Einstein* (New York: Simon and Schuster, 2007), 346.

of Hugh Everett[33] and the guided wave interpretation of David Bohm[34] and Louis de Broglie.[35] In both cases, the founder's intention was to ascribe some kind of real, physical meaning to the "ghostly" wave function that controls the particle's dynamic behavior as opposed to the Copenhagen interpretation that wave functions are purely mathematical abstractions that are devoid of any kind of physical meaning unto themselves. Today most scientists are very comfortable with the Copenhagen interpretation, and the field of physics considers these later interpretations to be more in the realm of philosophy than an essential and evolving part of the body of science.

However, what if Einstein was right—or at least what if he had a valid point in his criticisms? Perhaps Einstein was more right than he knew, when he remarked, "God doesn't play dice." Perhaps there is an opening here for God to enter the picture in the subatomic world as the final determining factor, fixing the outcome of all quantum mechanical measurements! Quantum mechanics is only capable of calculating the chances of various outcomes and not the outcomes themselves! Perhaps the determination of the final outcome is a role God reserves for himself. Perhaps seemingly random subatomic events are not just meaningless rolls of some abstract dice but are in reality choices made by God—and God alone—to communicate God's will in ways and for reasons that are impossible for us to fathom. Perhaps the origins of godly communication to the material universe lie deeply within the mystery of quantum uncertainty, which science has chosen arbitrarily to understand as a series of random events that are devoid of any kind of meaning by themselves.

A further word must be said at this point about the "God of the gaps" problem. In the early days of scientific development, mysteries abounded that early scientists could not explain using their elementary scientific methods. To explain away these gaps in their knowledge, many early

---

[33]   Hugh Everett (1930–1982) was an American physicist best known for his "many worlds" interpretation of quantum mechanics.

[34]   David Bohm (1917–1992) was a British physicist who is best known for his "guided wave" interpretation of quantum mechanics.

[35]   Louis de Broglie (1892–1987) was a French physicist and founding father of quantum mechanics. De Broglie's theory of matter waves is one of the principal pillars of the founding of quantum mechanics. He received the 1929 Nobel Prize in physics.

scientists invoked God as the likely cause. For instance, Kepler wrote that it was directly through the agency of God that the planets moved in their orbits around the sun. It was nearly two hundred years later when Newton finally closed out this gap using the mathematics of calculus contained in his mechanics. Later, as science advanced, other gaps were filled in with purely mechanical causes, much to the discredit of any theory that relied on divine intervention alone to fill these mysterious gaps in our understanding.

We want to state clearly and strongly that quantum mechanics is in no way a case of God of the gaps. Quantum mechanics cannot and will not ever be able to "fill in" any ignorance of an entity's properties in the way of deterministic physics. Albert Einstein tried for over thirty years and failed. Later, John Bell's inequality proved beyond a shadow of a doubt that quantum mechanics provides the maximum possible amount of information that can ever be known about a given subatomic system.

Einstein privately continued his fight to disprove the Copenhagen school's interpretation throughout his tenure at the Institute of Advanced Studies in Princeton, New Jersey, during the late 1930s, '40s, and '50s. After he emigrated to the United States, Einstein never again publicly debated Bohr on the meaning of quantum mechanics. His private battle culminated in what has become known as the EPR paper[36] in which Einstein explored what he called the "spooky action at a distance" that is associated with a pair of particles sharing a common beginning but traveling along different paths through space-time. Einstein intended this particular *gedanken* (thought in German) experiment to spotlight the illogic of a quantum situation enabling communications faster than the speed of light to occur. Such high-speed communications were expressly forbidden by the rules of special relativity. However, quantum entanglement, as this phenomenon has come to be called, has since been observed in the laboratory. Yet again, quantum mechanics has refused to do what is expected of it—that is to behave logically.

---

[36] The EPR paper (1951) was published by Albert Einstein and two colleagues at the Institute of Advanced Studies in Princeton on "entanglement" as his final attempt to cast doubt on the validity of the Copenhagen interpretation of quantum mechanics.

God Does Not Play Dice

Turning once again to theology, what were the core issues of the Bohr-Einstein debate? And what do these issues have to do with God? One of the few issues Bohr and Einstein did agree upon was that their debate should be limited to gaining a greater understanding of the meaning and philosophical implications of quantum mechanics. To this end, there was general agreement by both parties that it was important for them *not* to get bogged down in mathematical details but to maintain the debate at a purely philosophical level.

One of the important core issues was the probabilistic interpretation of the quantum wave function. A second core issue was the fundamental inability of any measurement to simultaneously determine all of a subatomic system's dynamic variables. This issue led naturally to the conclusion that it was the measurement process itself that confined reality to entities of the subatomic world.

This issue was a major stumbling block for Einstein. Bohr had dubbed his view of this principle "complementarities," and it soon became a pillar of the Copenhagen interpretation. Einstein, ever the master of wise but comical rejoinders, said about the complementarity issue, "Professor Bohr, do you mean to tell me the moon doesn't exist until I choose to look at it?"[37] Throughout these debates, Einstein raised some very reasonable questions; however, the basic problems still remained because quantum mechanics is rarely, if ever, reasonable!

Let's consider Einstein's comment, "God does not play dice." This much-quoted remark is very interesting because it consciously or unconsciously ties God to the outcome of a seemingly random series of events. If we are standing at a gaming table in Las Vegas and throw a pair of dice, why can't we predict the outcome? Imagine the profits we could make if only we could. Of course, dice were invented for just this purpose. The dice are constructed in such a way that makes the upturned numbers after a throw impossible for the players to predict. (Recall Julius Caesar's

---

[37] Manjit Kumar, *Quantum, Einstein, Bohr, and the Great Debate about the Nature of Reality* (New York: W. W. Norton and Company, 2008), 352.

famous words, "The die is cast."[38]) The differences in the initial conditions of the thrower's hands are very hard to duplicate. Otherwise skilled players might be able to control the outcome of their throws. It is also very hard to predict where and when the dice will strike the table. All of these factors are critical in determining the outcome of a throw. The conclusion must be that a throw of the dice always involves some knowledge that we simply do not and cannot possess. In fact, if we think about it carefully, we realize that this knowledge is beyond anyone's ability to possess.

The universe, as viewed through the laws of quantum mechanics, will not allow us to gain access to certain knowledge. If we can't access this knowledge, who can? Can anyone? Can God? Perhaps God alone knows the answer, and perhaps it is God who decides each outcome before we humans begin applying our statistics and calculating our probabilities. All we humans can do, no matter how skilled we are at statistics, is calculate the probabilities of outcomes and not the outcomes themselves. This is because the mathematics of statistics is a computational system that determines the behavior of entire populations even when the actions of the individuals who comprise the system are unknown and theoretically unknowable. Perhaps—as Einstein seemed to be suggesting—it is God who acts (or reserves the right to act) in each and every quantum mechanical measurement.

The Waves of the Wave-Particle Duality

There is a concept within quantum mechanics that is called the wave-particle duality. In its earliest and simplest form, wave-particle duality—referred to the recognition by Einstein, de Broglie, and the other scientists who were working in the earliest days of the quantum revolution—is the concept that all matter and energy have both a wave nature and a particle nature. Let's now dig a little deeper into this concept. According to the quantum mechanical understanding of wave-particle duality, the two complementary natures (wave and particle) are impossible to separate. This conclusion about quantum particles and waves is not unlike the

---

[38] Roman historian Suetornius attributed this statement to Julius Caesar as he and his army crossed the Rubicon River in northern Italy and marched fully armed on the city of Rome.

Christian theological understanding of Jesus's nature as formulated by the Council of Chalcedon in 451 CE.[39] Jesus was understood by the council to simultaneously be fully human and fully divine.

To go any further, we must first explore what is meant by a wave nature and what is meant by a particle nature. Let's start with a wave nature. Waves create a sense of *space*. They do so because, by mathematical necessity, all waves must occupy some region of space. Waves occupy space because all waves have a property called wavelength. Without wavelength, there can be no wave. Wavelength is something that can be calculated for every conceivable kind of wave. If a wave does not occupy sufficient space to measure at least one of its wavelengths, then in reality, the wave does not exist.

This concept is a basic property of all waves, including quantum mechanical wave functions. Waves can be traveling waves, moving in a direction determined by their wave vector, or they can be standing waves, forever marking time at one place in space. In an extreme form, certain traveling waves called *plane waves* mathematically have the ability to spread laterally over all the space lying within a plane perpendicular to the direction of their wave vector. Midway between plane waves and standing waves lies the general category of most waves that to some degree are confined in space and time, but they also travel in a certain direction at a certain velocity.

To a surprising degree, all waves are alike, and all waves have many, if not most, of their properties in common. The four most important of these wave properties are amplitude, wave vector, frequency, and phase. These common properties remain the same whether we are dealing with quantum mechanical wave functions, ocean waves, sound waves, radio waves, X-rays, or gamma rays.

All wave phenomena can be described by a branch of mathematics known as Fourier analysis,[40] which is named for the nineteenth-century

---

[39] The Council of Chalcedon (451 CE) was an important conference of the early Christian church that declared that Jesus was simultaneously fully human and fully divine.

[40] Joseph Fourier (1768–1830) was a French mathematician. Fourier is best known for his analysis of periodic mathematical functions. His approach was to generalize the description of any periodic functions as a series of sine and/or cosine functions, which bears his name.

French mathematician Jean Baptiste Joseph Fourier. In the mathematics of Fourier analysis, the description of a wave may be transformed back and forth between a space-time type of description and a frequency-wave vector description. Fourier analysis allows researchers in all branches of science to analyze arbitrary waveforms as a mathematical series of harmonically related sine and cosine waves. The term harmonically related refers to the property of these sine and cosine waves in which they have frequencies that are integer multiples of the lowest frequency sine and cosine waves in the series.

Perhaps the most important property that is shared by all waves is a phenomenon called interference. Much of the strangeness encountered in quantum mechanics originates from wave interference phenomena. In general, wave interference works like this: If two waves of the same frequency and amplitude are added together, the resultant wave (representing their sum) may be a wave of twice the amplitude of the individual waves. The sum of the waves may also have zero amplitude and cease to exist altogether. Finally, the sum may be anything in between two times the amplitude of the individual waves and zero depending on the relative phase (orientation) of the two waves! When this phasing effect occurs during interference, it is called coherence.

Phase is the scientific term for the relative orientation that exists between two waves of the same frequency and wave vector. Phase, like sections of a circle, is measured in degrees. Because of coherence, two identical waves always possess the ability to cancel each other out or alternately raise their sum to new heights depending on the relative phase orientation of the two waves.

It is this wave cancellation phenomenon that explains many of the unusual quantum mechanical behaviors encountered in the subatomic world. Examples are the gaps and discontinuities that occur in energy bands (i.e., quantum leaps or jumps) of atoms and molecules. The energy gaps in these bands are regions where certain energies (or momenta) are simply not allowed to exist (i.e., the wave functions associated with these disallowed regions have literally canceled each other out to nonexistence).

Of course, in the case of quantum wave functions, the question of "what is actually waving" always comes up. Is there a medium associated with the quantum wave function? With ocean waves, the medium is water; with sound waves the medium is air that is experiencing wavelike pressure fronts; and with radio waves, the medium is the electric and the magnetic fields that are doing the waving. In the case of electromagnetic waves, such as light, a key experiment was conducted about one hundred years ago by Michelson and Morley[41] that disproved the existence of any so-called "ether" medium that had been thought by many scientists to be responsible for radio waves and light propagation.

But quantum mechanical wave functions are also very different from other wave phenomena. The amplitude of a quantum mechanical wave function is not a physically measurable property; in fact, this amplitude is what is mathematically called a complex number. Complex numbers do not exist in the ordinary physical reality of the real number line. Therefore, the complex amplitudes of quantum wave functions do *not* have the same kind of physical reality as do the measurable amplitudes of water waves, sound wave, and radio waves. However, the key understanding here is that the probabilistic interpretation that is ascribed to the quantum wave functions is associated with a statistical chance of a certain entity having a certain physical property at a specific location in space and time.

For lack of any better way to describe it, we might say that for quantum mechanics, it is existence itself that is waving. But is the quantum mechanical wave function real? This is a very difficult question to answer, and the answer depends critically on what you mean by real. Perhaps the wave function is simply a mathematical abstraction whose role is to act as a computational tool to facilitate the calculations of subatomic phenomena. On the other hand, could the wave function have an interpretation that is in some way related to physical reality? This is a bedrock kind of question with no easy answers, or perhaps with no answer at all. It is the kind of question where scientific truth and religious truth meet.

Consider this type of question from the religious perspective for a moment. St. Thomas Aquinas,[42] in his *Summa Theologica*, stated his belief that ultimately the one who we call God is existence itself. Perhaps we

---

[41]   Albert Michelson (1852–1931) and Edward Morley (1838–1923) were American physicists who jointly performed an experiment that disproved the existence of an all-encompassing ether that had been thought to serve as a medium for the propagation of light waves.

[42]   St. Thomas Aquinas (1225–1274) was an Italian theologian and the author of the *Summa Theologica*, which was the most complete work of Christian theology and belief produced up to that point in the history of the church.

could think of the quantum mechanical wave function as a channel or approach through which God reaches from his unmanifested reality into our material universe to create finite and temporary manifestations of God's will. God's reach from this point of view is to the most elemental level of our material existence, namely the finest-grain realm of the subatomic world.

In Annie Dillard's book *Pilgrim at Tinker Creek*, she quotes Arthur Eddington as saying that being forced to drop causality as the result of quantum uncertainities leaves us with no clear distinction between the natural and the supernatural.[43] If God alone can control the outcome of individual events that quantum wave functions are statistically approximating, then perhaps we can come to recognize that our universe is most assuredly graced and directed by God's divine presence at its most fundamental level.

The Particles of the Wave-Particle Duality

Let us return to our discussion of waves and particles. Next we consider the nature of the particle side of the wave-particle duality. Particles are in all respects the exact opposite of waves. While waves are distributed throughout all of space-time, particles have all their properties concentrated at a single point in space-time. This point may move about in due course as dictated by the particle's dynamic behavior, but the full complement of the particle's properties move right along with the particle.

Particles have well-defined and completely measurable dynamic properties (i.e., position, momentum, energy, and timing). Particles interact with each other in much the same way as macroscopic bodies. We may think of particles as being subatomic billiard balls, and we would not be wrong. Their interactions always obey the laws of conservation of energy and conservation of momentum. There are no mysterious uncertainties, no probabilities, no statistics, and in general no weirdness. Particles are everyday, down-to-earth, rational, classical entities. What you see is what you get with a particle. There are no "it depends" when dealing with particles. Particles are always very dependable and straightforward.

---

43   Annie Dillard, *Pilgrim at Tinker Creek* (New York: HarperCollins, 1974), 202.

Particles live in the same physical landscape that was envisioned by Isaac Newton over two hundred years ago. From the point of view of quantum mechanics, Niels Bohr and the Copenhagen school taught that all quantum measurements are classical measurements. This means that at the moment of measurement, the wave function has collapsed and a quantum measurement becomes the measurement of particles. We will see this concept demonstrated in the two measurement examples that follow.

The Single- and Double-Slit *Gedanken* Experiments

During the Einstein-Bohr debates of the 1920s and 1930s, much of the debate revolved about questions of when a quantum entity had wave-like behaviors and when it had particle-like behaviors. Is the assumption of an indivisible wave-particle duality really true, or under special circumstances, can there be purely particle behaviors and/or purely wave behaviors? Can these unique behaviors be separated or distilled out in the course of running a properly configured experiment?

In an attempt to answer these difficult questions, two very cleverly configured *Gedanken* ("thought" in German) experiments were proposed and hotly debated at a number of physics conferences. The first was called the single-slit experiment, and the second was called the double-slit experiment. As we shall see, both of these Gedanken experiments in the final analysis backfired relative to the original intentions of their creators, but they succeeded in yielding insights into the nature of quantum mechanics that no one had expected. This pattern of debate was so typical of the development of quantum mechanics. Many logical "what if" questions were being asked by sincere investigators as they attempted to control the seemly illogical direction in which quantum mechanics was heading. The results were often a great surprise to the questioner and most often served only to deepen the mystery rather than resolve it.

The single-slit and double-slit Gedanken experiments, perhaps more than any others, served to bring into sharp focus the many seemingly mysterious aspects of quantum mechanics. Many physicists, then and now, feel that by gaining a truly in-depth understanding of the single—and double-slit experiments, one can gain a deeper appreciation for what quantum mechanics is trying to tell us, as weird as it may seem. In the case of both Gedanken experiments, the final conclusions tended to confirm an already gathering understanding that the Bohr-Heisenberg Copenhagen team was on the right track all along in their understanding of quantum mechanics.

The Single-Slit Gedanken Experiment

First consider the single-slit experiment. Both the single—and double-slit experiments are based on similar experiments from the field of classical geometric optics. An illustration of the single-slit experiment is shown on page 30. The single-slit experiment is mentally performed as follows.

A beam of electrons (or some other kind of subatomic particle) is directed toward a very narrow slit in a metal plate that otherwise (without the slit) would stop the electron beam from penetrating any further. On the far side of the plate, there is a photographic film that has the ability to permanently record the point of impact of an electron that passes through the slit. We assume there is some mechanism for making the slit wider or narrower, as the experimenter chooses. When the slit is very narrow, the electrons are confined in the Y direction (along the plane of the metal plate), a condition that corresponds to a more particle-like behavior (i.e., the concentration of all of the electron's properties at a single point in space time), at least in the Y direction. As the slit is closed down to nearly nothing, any electrons that pass though the slit must have a Y position that is known very accurately, and therefore the electron has become very particle-like. By configuring the experiment in this way, the experimenter has forced the electrons to behave in a semi-particle-like way, at least in the Y direction. If we accurately knew the electron's Y-directed momentum, at least in principle this pair of dynamic variables would be accurately known at the same time, leading us to conclude that at least in the Y direction, these electrons are acting purely as particles.

However, there is a problem with drawing such a simple conclusion. Instead of the electrons flying straight through the slit, as we might imagine from everyday experience, the real-life situation is that the electrons tend to fan out on the far side of the slit, just as light rays entering a pinhole camera would. Because of the electrons' wavelike behavior, they are forced to obey all of the laws of geometric optics. What really happens here is that the electron's wave nature is reasserting itself by causing a behavior that conforms more or less to the laws of geometric optics. Since the wave nature of the electrons on the far side of the slit cannot be ignored (as the slit is made narrower), the fan-out of individual electrons increases until the point is reached when the slit is nearly closed and the fan-out angle becomes nearly 180 degrees.

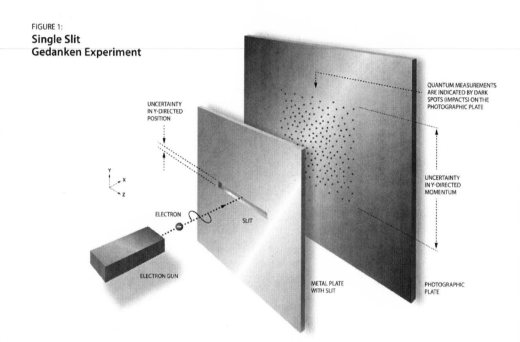

FIGURE 1:
ILLUSTRATION OF THE SINGLE-SLIT
GEDANKEN EXPERIMENT

Momentum, as we know from Newtonian physics, is a vector quantity and always has an associated direction. Since the fan-out angle is associated with uncertainty in the electron's Y-directed momentum, the more the electron's Y-position is confined by closing the slit (in the attempt to force nearly purely particle-like behavior), the greater the fan-out of the electrons as they pass through the slit. Therefore, the harder we try to reduce the uncertainty in the electron's position, the greater the uncertainty in the electron's momentum!

We are forced to conclude, much to the surprise of the experiment's original creators, that the single slit-experiment proves the fundamental impossibility of any quantum entity ever being either a pure wave or pure particle. At the same time, the single-slit experiment provides a perfect demonstration of how the Heisenberg[44] uncertainty principle works. As we would expect, based on the Heisenberg's uncertainty principle,[45] reduced uncertainty in an electron's position (which is created as the slit is closed down) must always be accompanied by an associated increase in the uncertainty of the electron's momentum, which is exactly the conclusion of the single-slit experiment!

This dramatic conclusion was exactly the opposite of what was expected by the scientists who proposed the single-slit experiment in their attempt to disprove the reality of wave-particle duality and Heisenberg's uncertainty principle. However, contrary to the hope of its originators, this much-debated Gedanken experiment has in the final analysis provided strong evidence for the reality of both wave-particle duality and the Heisenberg uncertainty principle!

---

[44] Werner Heisenberg (1901–1976) was a German physicist and a chief contributor to the Copenhagen interpretation of quantum mechanics. Heisenberg was responsible for the development of the matrix mechanics formalism of quantum mechanics and the uncertainty principle that bears his name.

[45] The uncertainty principle is the principle by which certain pairs of dynamic variables in a quantum mechanical system cannot be measured simultaneously with complete accuracy. As the measurement accuracy for one of the variables is increased, the measurement accuracy for the other variable *must* decrease.

The Double-Slit Gedanken Experiment

As if the single-slit experiment had not caused enough egg on the faces of the physicists who opposed the Copenhagen interpretation, a more-sophisticated Gedanken experiment called the double-slit experiment was next proposed in an attempt to debunk the general quantum weirdness many scientists, including Einstein, found so unacceptable. An illustration of the double-slit experiment is shown on page 33. The setup for this experiment was very similar to that of the single-slit experiment, except that in the case of the double-slit experiment, the metal plate contained two slits instead of one slit. The distance between the slits could be controlled by the experimenter, as could the width of each slit. Again, as in the single-slit experiment, a photographic film is mounted on the far side of the metal plate (i.e., away from the electron beam source). This photographic film records the point of impact of all electrons that pass through the plate's two slits.

In the double-slit experiment, when an electron beam is directed toward the middle of the solid metal plate between the two slits, no electron should be able to get through the slits based on classical Newtonian physics. The photographic plate on the far side of the metal plate should remain unexposed except for an occasional misdirected electron moving at a very large angle relative to the centerline of the main beam. These occasional maverick electrons are expected to record their presence on the photographic film directly behind one of the two slits. We would expect to find photographic film smudges slowly piling up directly behind each of the two slits. However, we would not expect to find any photographic film smudge records between the two slits.

**Double Slit
Gedanken Experiment**

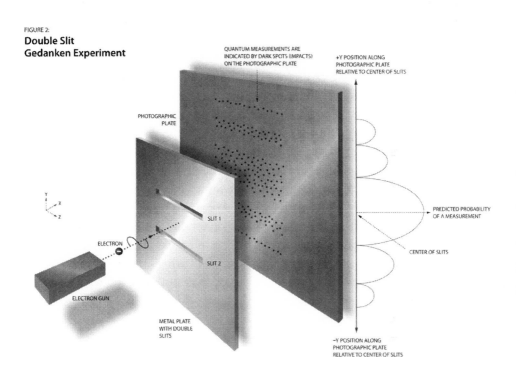

QUANTUM MEASUREMENTS ARE
INDICATED BY DARK SPOTS (IMPACTS)
ON THE PHOTOGRAPHIC PLATE

+Y POSITION ALONG
PHOTOGRAPHIC PLATE
RELATIVE TO CENTER OF SLITS

PHOTOGRAPHIC
PLATE

Y

X

Z

PREDICTED PROBABILITY
OF A MEASUREMENT

SLIT 1

CENTER OF SLITS

ELECTRON

SLIT 2

ELECTRON GUN

METAL PLATE
WITH DOUBLE
SLITS

–Y POSITION ALONG
PHOTOGRAPHIC PLATE
RELATIVE TO CENTER OF SLITS

FIGURE 2:
ILLUSTRATION OF THE DOUBLE-SLIT
GEDANKEN EXPERIMENT

Strangely, and this is so typical of quantum mechanics, the expected behavior based on our intuition from everyday life experience is not what really happens. The reality is that as a direct result of the electron's wave nature, each individual electron in the beam presents a nearly plane wave front at the metallic plate. The electron's wave front strikes the two slits in the metal plate simultaneously. This means that the electron's wave front strikes each of the slits with equal phase and equal amplitude. As the wave front passes through the two slits, each electron's wave behavior becomes a superposition of two waves, one wave from the upper slit and one wave from the lower slit. The electron's wave superpositions pass through both slits, and just as in the case of the single-slit experiment, each slit acts as a miniature pinhole camera, fanning out each wave front as it moves toward the photograph film. (The fan-out angle depends on the width of the slits, as in the case of the single-slit experiment.)

Because of coherent wave interference effects, the probability of each electron causing a smudge on the film alternates in a series of Y-directed smudge bands on the film as the electron wave superpositions alternately interfere constructively and destructively with each other. (The coherent wave interference effects are caused by the phase differences between each of the wave superpositions as they arrive simultaneously at some point along the Y-axis of the photographic film.) Of course, the photographic smudges only build up clarity as more and more electrons pass through the slits, fulfilling the large population requirements of the probabilistic nature of quantum measurements.

These smudge bands are arranged vertically on the photographic film and define regions of high probability of impact and low probability of impact. Within this interference pattern, the two wave superpositions alternately combine constructively and destructively (as their wave functions progressively move in and out of phase along the photographic film).

For the in-phase regions, there is significant dark smudging, and for the out-of-phase regions, there is little or no smudging. Therefore, as more electrons arrive at the photographic film, these bands gradually fill in. This yields the very surprising and highly unintuitive result that the darkest smudging (indicating the largest number of electron impacts at the photographic plate) occurs directly between the two slits, since this is the point where the two wave superpositions add up constructively in phase over the widest possible area of the photographic film! However, it is this very region of the film that is located exactly where there should be no electron impacts at all if we were dealing with were electrons with a purely particle nature.

The double-slit experiment proves that electrons can literally pass through walls. Our everyday common sense would expect purely particle-like electrons to behave just like hard little billiard balls bouncing off the metal plate that lies directly between the slits. However, what we really observe (if we were to run this Gedanken experiment in the laboratory, which in fact has been done many times[46]) is the surprising result that, because of their wave natures, electrons, behaving according to the rules of quantum mechanics, are capable of literally passing through walls! However, once they have passed through the metal wall barrier, individual electrons have no trouble regaining their particle-like natures to record their presence at the proper point along the photographic film. These particle-like electrons must still obey all of the rules of conservation of energy and conservation of momentum as they interact with the silver atoms in the photographic film.

[46]   Jim Al-Khalili, *Quantum: A Guide for the Perplexed* (United Kingdom: Weidenfeld and Nicolson, 2003), 24–25.

By working through the details of the double-slit Gedanken experiment, we come to appreciate how really strange and unintuitive the behaviors of quantum entities are. These quantum entities (such as electrons) seem to have a ghost-like ability to divide themselves in half to get around barriers and then reconnect their two halves on the other side of the barrier. This phenomenon, called tunneling, is very real and serves as the basis of operation for many modern microelectronic devices, such as tunnel diodes.[47] Tunneling also serves as the mechanism for ejecting alpha particle radiation from the nucleus of some radioactive atoms.[48] These improved theories of the atomic nucleus led naturally to an investigation and cataloging of the radioactive isotopes of many elements, which culminated in the discovery of nuclear fission by Lisa Meiter[49] in the

---

[47] Tunnel diodes are an electronic device that operates on the principle of quantum tunneling.

[48] Alpha particle radioactive decay is a type of naturally occurring radioactivity in which helium nuclei are ejected from an atom's nucleus as a direct result of quantum tunneling.

[49] Lisa Meitner (1878–1968) was an Austrian and Swedish physicist born in Vienna, Austria, into a well-to-do Jewish family. Lisa studied physics at the University of Vienna at a time when women were discouraged from entering this field. She went on to receive a PhD from the University of Vienna under the direction of Ludwig Boltzmann. She later moved to the Kaiser-Wilhelm Institute in Berlin Germany, where she worked under Max Planck as his assistant. She worked with the chemist Otto Hahn, and together they discovered several new isotopes and published papers on beta radiation. During this period, Lisa worked without salary as a guest in the department of radiochemistry. However, because of the uniqueness of her work, by 1926 Meitner had become the first woman in Germany to become a full professor of physics at the University of Berlin. In the 1930s she turned her attention to nuclear physics and chemistry. She and Hahn competed with research groups in Italy, France, and England to create new and heavier elements with neutron bombardment. In 1933 Adolf Hitler came to power, and many Jewish scientists in Germany lost their jobs and were forced to leave the country. Meitner felt she was protected by her Austrian citizenship. However, by 1939, Austria had been annexed by Germany and Meitner was forced to flee, first to the Netherlands and later to Sweden. However, she was able to stay in contact with Hahn back in Germany and later formed a partnership with Neils Bohr (in Copenhagen), who often traveled to Stockholm. During

late 1930s. However, it is very important to understand that the strange behaviors encountered in the double-silt experiment are very typical of the behaviors of electrons and other particles within the atoms and molecules that are all around us and are in fact a part of us. It is quantum mechanics alone that serves to define the reality of the subatomic world.

The ultimate meaning of the double-slit experiment lies in its ability to provide us with a mental picture of how quantum entities are capable of shifting seamlessly back and forth between their particle nature and their wave nature, depending upon circumstances. However, a very important conceptual understanding to be gained from this experiment is that all quantum entities remain wave-like as long as they don't interact (that is, exchange energy and momentum as particles) with the measurement apparatus (i.e., the photographic film's silver halide crystals).

---

this period, Meitner began to realize that some uranium atoms would literally split in half (fission) when exposed to neutron radiation of a certain kind. Back in Germany, Hahn did chemical experiments that conclusively proved that light elements, such as barium, were present after the disintegration of uranium atoms that had been bombarded by neutrons. Using the "liquid drop" model of the atomic nucleus, Meitner was able to explain theoretically how the nucleus of an atom could be split into smaller, lighter parts along with the production of several more neutrons and significant amounts of free energy. The energy production was a direct result of the conversion of some nuclear matter into energy according to Einstein's famous equation $E=mc^2$. Because of the political situation at this time, it was impossible for Meitner and Hahn to publish their results jointly. Hahn and a colleague named Strassman published his results in 1938, while Meitner and her nephew, Otto Frisch, published a different paper in 1939 interpreting their results as proof of the reality of nuclear fission. Meitner recognized the potential for a chain reaction with enormous releases of energy. Such a chain reaction could lead to terribly destructive weapons or to electrical power generation. Several of the Jewish scientists who fled Nazi Germany at this time carried word of Meitner's fission chain reaction to America, where in time the Manhattan atomic weapons project was the result. Hahn received the Nobel Prize in chemistry in 1945 for the discovery of nuclear fission. Meitner's contribution was unfairly ignored by the Nobel committee, and she never received a Nobel Prize. However in 1966, Hahn, Strassmann, and Meitner jointly received the Enrico Fermi award in the USA. A new element, Meitnerium 109, was later named in her honor.

Going one step further, when a measurement does take place, the electron's particle nature takes over, and an exchange of energy and momentum takes place that must obey all of the conservation rules, locating the electron's point of impact at a particular point in space and time (i.e., somewhere on the photographic plate). Bohr referred to the moment when this measurement event occurred as the collapse of the wave function. From the point of view of the Copenhagen interpretation, instantaneous wave function collapse must always precede a measurement. In Bohr's opinion, quantum measurements always obey the rules of classical physics (i.e., the physics of Newton), requiring the entity's particle nature to manifest at the moment of measurement. It is this "coming into its particle nature" that allows the conservation rules' bookkeeping to be worked out in detail, as it must. However, the values of the entity's dynamic variables at the exact moment of measurement can only be predicted in a statistical way based on the details of electron's wave function superposition at the moment of wave function collapse. Perhaps it is at this critical moment of wave function collapse when divine action enters our universe.

Although quantum mechanics offers science the key to understanding the subatomic world, such understanding comes at a mighty price: the end for all time of scientific determinism. This was and still is a major problem for many scientists. However, science's loss is religion's gain. It is the very unpredictability of quantum mechanical uncertainty that destroys determinism and simultaneously opens the door for communication between God and the natural universe. Perhaps it is quantum phenomena that serve to constantly connect God to his creations, including ourselves. In the next few chapters, we explore the implications of these statements.

CHAPTER 3

THE GREAT PARADOX

Paradox

Many branches of science are able to advance themselves by the process of resolving paradoxes. This is especially true of the field of physics, where progress forward has been made again and again as the direct result of facing up to various paradoxes as they have arisen. It is the greater, more-fundamental understanding that comes out of resolving of a paradox that pushes a field forward. A paradox is a seeming conflict between two conceptual ideas, both of which you believe to be true.

Stanford physicist Leonard Susskind, in his book *The Black Hole War*, presented a list of the most notable paradoxes from the history of physics. The following is Susskind's physics paradox list.[50] The important point here is that in each case, it is the solving of the paradox that moved the field of physics forward toward an ever-deepening understanding of our natural world!

1. Galileo (and even more so Newton) was greatly distressed by the commonly accepted wisdom of their day (and going all the way back to the ancient Greeks) that laws governing the celestial phenomena of planets, stars, comets, and galaxies had nothing at all to do with the laws governing the terrestrial phenomena of objects here on the earth. It was Galileo who came up with the critical Gedanken experiment. (We are sure that Galileo would have used

50 Leonard Susskind, *The Black Hole War* (New York: Back Bay Books, 2008), 202–10.

an Italian name instead of Gedanken, but today, there is so much tradition in physics for using the German name Gedanken that we will continue its use here.) Galileo imagined himself standing on the top of a high mountain. He picked up a stone and threw it a small distance from himself, and it fell to the ground. Next he picked up another stone and threw it a good deal harder, and it flew off the mountaintop but landed in the valley below. Next Galileo imagined that he possessed superhuman strength so that when he threw the next stone, it traveled sufficiently fast to never return to earth but in fact broke the gravitational bonds of the earth and went into orbit about the earth. The paradox is that what was once a terrestrial object (a rock) has now become a celestial object (a rock orbiting the earth). Galileo and later Newton could not accept that this rock was obeying any different laws in orbit than it had previously obeyed on earth. Newton solved this paradox once and for all with his discovery of classical mechanics[51] and his invention of the mathematics of calculus. Newton's mechanics successfully demonstrated that planets, comets, and stars all obey the same laws of mechanics as earthbound objects.

2. Austrian physicist Ludwig Boltzmann[52] was greatly troubled by a paradox that existed between Newton's mechanics, which is completely time reversible (according to Newton's mechanics, one can always run an event backward in time if one chooses, with no ill effects), and thermodynamic entropy, which implies

---

[51] Classical mechanics is a system of mechanics first described by Newton in the late seventeenth century. In classical mechanics, the sum of the forces on any mass is equated to the mass times the mass's acceleration. Expressed mathematically, this concept is summed up by the equation F=ma, where F is the sum of the forces, m is the mass, and a is the mass's acceleration.

[52] Ludwig Boltzmann (1844–1906) was an Austrian physicist who was best known his work on statistical mechanics. Boltzmann famously express the entropy of a many-bodied thermodynamic system (such as a gas) as Boltzmann's constant (k=1.38E-23 joules/degree Kelvin) times the logarithm of the number of ways the system can arranged (complexions). In equation form, Boltzmann's entropy is expressed as S=k log (N), where N is the number of "complexions" in which the system can be arranged.

that all events must have a required forward direction in time (entropy always increases as time passes and cannot be reversed to run time backward). Boltzmann solved the paradox by using atomic theory (which was not universally accepted in Boltzmann's time) and techniques from statistical analysis to determine that because all objects are made up of incredibly large groups of atoms, these groups are unlikely to ever reassemble in their present form because they carry no memory of the past, rendering time reversibility an impossibility. Boltzmann's insight can be summed up in the children's rhyme, "Humpty Dumpty sat on a wall. Humpty Dumpty had a great fall. All the king's horses and all the king's men couldn't put Humpty Dumpty back together again."

3. Albert Einstein (when he was an unknown patent clerk in Bern, Switzerland, who, in 1905, wrote several papers that literally stood the field of physics on its head!) discovered the principles of special relativity[53] while trying to resolve the paradox of how the laws of physics (i.e., Newton's mechanics) could be acting in the same way for all constant velocity frames of reference and James Clerk Maxwell's discovery that the speed of light is a universal constant, independent of the observer's velocity. Einstein's Gedanken experiment was imagining himself riding in a train car that was moving at the speed of light. The question Einstein asked himself was what would he see if a light beam was moving along beside the train? At first he wondered if he would see a standing wave frozen into a stationary position. Later he realized this would be nonsense, because light, whose speed is a universal constant, would always be moving at the speed of light inside the train, no matter how fast the train was going. This insight led Einstein to ask important questions about what is meant by simultaneous

---

[53] The concept of special relativity was presented in one of Einstein's 1905 papers. Based on James Clerk Maxwell's classical electrodynamics, Einstein concluded that the speed of light is the same of all bodies, moving or at rest, no matter what their velocity. This conclusion led to some very interesting paradoxes when a moving system is observed by a stationary observer. As the moving system approaches the speed of light, it will appear to a stationary observer to be shrinking in size and to have a timekeeping that is slowing down.

events and perceived changes in lengths and time passages for various observers in different reference frames moving at different velocities. Out of Einstein's paradox, and its solution, the insight of special relativity was born.

4. Again in 1905, Albert Einstein solved a second paradox called the ultraviolet catastrophe. The problem was that according to classical physics, each wavelength of electromagnetic radiation (i.e., radio waves, light waves, and X rays) should contain an equal amount of energy. The problem here was that at very short wavelengths, the amount of energy being contributed to a sum of total of energy remained the same. The Gedanken experiment was to imagine a heated box containing electromagnetic radiation at many wavelengths. It was well known from the principle of blackbody radiation that in this situation, each wavelength would have an energy that depended on the box's temperature alone. Since there was no limit to how short the radiation's wavelength could be, the energy at progressively shorter wavelengths just kept adding up (i.e., at ultraviolet wavelengths and shorter) until it became infinite and destroyed the box. Clearly there was no way for the energy inside the box to be infinite, and for that matter, there was no reason to think that the box was in any danger of being destroyed. But what was happening? Einstein solved the paradox by proposing that light (and all electromagnetic radiation) was composed of particle-like quanta of energy called photons, each of which has an energy proportional to the radiation's frequency (which is equal to the speed of light divided by wavelength). Einstein applied Boltzmann's statistical method to analyze the photons in the heated box and discovered there being fewer and fewer photons at shorter wavelengths, until there were none at all. Since there are no longer any photons at shorter wavelengths (which is where the problem was), the total energy in the box no longer becomes infinite, and the paradox is resolved! The quantization of light into particle-like photons does not invalidate Maxwell's wave theory of light but actually led to the recognition that a wave-particle duality must exist in the case of light. A few years later, Louis de Broglie showed how a similar wave-particle duality also exists for matter, and quantum mechanics was born!

5. By 1916 Einstein was no longer just an unknown patent clerk. He had already achieved a degree of fame in the world of physics based on the success of his 1905 papers. However, it was in 1916 that a still-young Einstein produced his magnum opus, the theory of general relativity.[54] The paradox that generated general relativity was Einstein's recognition that with Newton's mechanics, there are two distinct definitions of mass. The first definition relates to a mass's inertial property (a law relating force to acceleration). The second definition of mass accounts for the gravitational attraction of one mass to another mass. Einstein solved this paradox by boldly proposing that the effects of gravity and acceleration are indistinguishable from each other. Where this path led was literally to the development of a theory for the entire cosmos and ultimately, as we have already seen, to the big bang theory of an expanding universe.

The Big Questions

If you were to stop people at random on the street and ask their opinions on the most profound questions facing humanity today, you would probably get a wide range of answers. Some people would say, "Can the world ever live in peace?" Others would say, "Will there ever be real justice and equality for all of the world's people?" Some would ask, "Can the earth's environment be saved before it is too late?" Still others would

---

[54] General relativity is a grand and universal conceptual system that was developed by Einstein in 1915. General relativity grew out of resolving a paradox of why Newton's mechanics contains two different definitions of mass (gravitational and inertial). Einstein concluded that both definitions of mass had to be equivalent, leading to the concept that mass can be thought of most accurately as distortions in space that lead to spatial gradients that cause accelerations. In special relativity, even light beams are bent when passing close to a large mass (such as a star). The concept of general relativity may be expressed in equation form as: Ru-(1/2) gu R = 8 π G Tu, where Ru is the Ricci curvature tensor, gu is a metric of space-time, R is the scalar curvature, G is Newton's constant, and Tu is the energy-momentum tensor.

ask, "Is there life elsewhere in the universe?" And finally some would ask, "Does God really exist?"

None of these questions, although they are all worthy and profound in their own right, really gets to the heart of the matter in the search for a unified truth. The authors of this book believe the most-profound question facing seekers of a unified truth (of course, assuming that God exists) is, "How does God communicate with the material universe in a way that is consistent with both religious truth and scientific truth?" Remember, God may indeed exist, but we would have no way of knowing about God's existence if God would not, or could not, communicate in some way with the material universe.

Perhaps an un-communicating God is the point of view held by deists, but for theists (who comprise about 99 percent of the world's religious people), communications between God and our material universe (and everyone in it) is absolutely essential. Without communications from God, there can be *no* religious truth. We would all become like the people in Plato's cave, unaware of the existence of the wider world of God. But by assuming the existence of godly communications, we open up a challenge to scientific truth because science must somehow understand how such godly communications could occur in a way that does not violate the laws of science.

If we take the point of view that God can suspend the laws of science, as necessary, to facilitate communications with our universe, then no unity of scientific and religious truth is possible because the laws of science are no longer universally true. Scientific truth would have been *subordinated* to religious truth! The only hope for truly unifying the truth is to answer, to the best of our ability, the question of how God achieves communications with our material universe consistent with the laws of science! Only by providing credible and believable answers that are acceptable to both science and religion can real progress be made toward a truly unified truth. In the remainder of this chapter, we will examine what scientific hurdles must be bridged to answer this question in the affirmative. In the following chapter, we will explore credible solutions to these hurdles and in the process point the way to a more-profound understanding of ourselves and our relationship with God.

We begin our quest for the unified truth in a branch of science that has struck terror into the hearts of students since the beginning of scientific education: thermodynamics! Thermodynamics seems like an unlikely place to start our quest for understanding the mechanics of godly communications. It turns out there is a deep and profound connection between thermodynamics and communications that will serve to point us in the right direction so we can ask the right questions.

All the King's Horses and All the King's Men Couldn't Put Humpty Dumpty Together Again

During the Industrial Revolution, the steam engine engineers of the late eighteenth century were busily engaged in designing bigger, better, more powerful, and more efficient steam engines. These engines were of profound importance as the prime movers of the industrial revolution. Steam engines powered the expanding factories, mills, mines, steamships, and railroads of the time. Many of these engineers began asking very fundamental questions concerning the operation of heat engines in general. The answers to these questions ultimately led to a vastly expanded understanding of our natural world. Today, this knowledge is encapsulated in the science of thermodynamics. Science now recognizes that thermodynamics is one of the most important underpinnings of physics, chemistry, and all of the branches of engineering. Like Newton's mechanics before it, thermodynamics has come, over the years, to be succinctly stated in just a few profound and fundamental laws.

The first law of thermodynamics is simply a statement of conservation of energy. The first law is often stated as: The sum of the energy inputs to a closed system must exactly equal the sum of the energy outputs from the same closed system (minus any energy that is stored within the system). Another way to state the first law of thermodynamics is to say that energy can neither be created nor destroyed but only transformed. The first law of thermodynamics is a scientific energy version of the credits and debits rules of accounting. In today's world, the first law of thermodynamics seems to most of us like a no brainier, but in the late eighteenth century, the concept of the first law was very difficult for people to accept.

Reaching for the big picture, we can say that within the complete generality of cosmology, the first law of thermodynamics can be stated as: The total energy of the universe is an unchanging constant! This means that every scrap of energy that is present in today's universe was also present at the moment of creation.

45

The second law of thermodynamics is not quite so simple to state or to understand. This is not entirely true, because computationally the second law is just as easy to deal with as the first law, but conceptually the second law of thermodynamics can really stretch your imagination. Generations of science students can remember their experience of suffering through "thermodamnamics!" When they say this, they are really talking about the second law. We believe that at some level, all who choose to grapple with the second law are forced to struggle with it throughout their entire lives. It is truly a long-term project to really understand what the second law is saying. As an added complication, there are at least three equally valid ways of expressing the second law. All three expressions of the second law initially sound deceptively simple—but don't be deceived; there is nothing simple about the second law of thermodynamics!

The first way of stating the second law is to say that any kind of physical action will always raise the entropy of the closed system in which the action takes place. At the most profound level of generalization, this conclusion must include the whole universe, which, of course, is the largest closed system we can conceive of. We will return to the meaning of entropy in a moment.

A second statement of the second law is that heat energy always flows from a hot body to a cooler body. This statement is really saying that on its own, heat energy can never flow from a cold object to a hot object (i.e., we all know from our life experience that it takes a certain amount of energy input, in the form of electricity, to run a refrigerator). A third statement of the second law says that it is not possible to build a perpetual motion machine. Any real mechanism requires some amount of continuous energy input (though it may be very small) to keep it running. If we build some mechanism (say a clock) within a closed system and start it running, with no ongoing energy input, the mechanism will eventually slow down and stop. A clock's pendulum, when the clock's weights have all run down, is a good example of what we are talking about.

It turns out that all statements of the second law are really saying the same thing in their own particular way. Each statement is like a facet of a beautiful gem. The third statement of the second law seems intuitively true, much to the sorrow of some inventors (i.e., there really is no free lunch in thermodynamics). The second statement seems to agree with our everyday experience (have you ever seen an ice cube get colder on a hot day?). But the first statement stops most of us dead in our tracks.

Just for openers, what is entropy? The quick answer is that entropy is a physical property of any material system (just like temperature, pressure, or volume) that can easily be calculated by dividing the energy input to the system by the system's temperature.

The following simple example will demonstrate why the entropy of a closed system always increases. Assume that our system is a glass of hot water and an ice cube. Now we take action by placing the ice cube into the glass of hot water. What happens? The answer, of course, is that the ice cube melts! But let's dig a little deeper and examine the heat energy flows within this system. In order to melt the ice cube, energy must flow out of the hot water and into the ice.

The following simple calculation (in words) will convince us that there is no way for the entropy of this system to decrease under any condition. By applying the first law of thermodynamics (conservation of energy) to this system, we know that the energy flowing out of the hot water must exactly equal the energy flowing into the ice cube. The energy flowing out of the hot water *decreases* the entropy of the hot water by an amount equal to the energy flowing out of the hot water (and into the ice cube) divided by the hot water's temperature. Similarly, the energy flowing into the ice cube *increases* the ice cube's entropy by the same energy (i.e., the energy flowing out of the hot water and into the ice cube) divided by the temperature of the ice cube. Since the temperature of the ice cube (about 32 degrees Fahrenheit) is always significantly less than the temperature of the hot water, it is a simple mathematical exercise to show that the total entropy of our hot water-ice system *must always increase,* no matter what the temperatures of the hot water or the ice cube! (Try it for yourself by guessing some numbers, remembering that the temperature of boiling water is 212 degrees Fahrenheit, and guess the energy transfer to be ten calories.)

In the next chapter we will associate entropy with lost information. In the case of the melting ice cube, we can understand the entropy increase to be associated with the loss of the information that was contained in the order associated with the ice cube's crystal structure. The ice cube melts, and its water molecules assume a more random (chaotic) configuration as liquid water. This seeming everyday result only becomes remarkable when we realize that all other similar actions taking place throughout the universe (from colliding galaxies to the jet engines of jumbo jets) are working constantly to increase the entropy of the universe! Therefore, whatever entropy is, the universe of the future is assured of having a lot more of it than the universe of the past. Can you think of anything else that always increases? We can't (with the possible exception of taxes, of course!).

There is also an implied time element to this process, which needs to be examined more closely. Let us return to our hot water-ice cube system. As the ice melts, the cold water of the frozen ice combines with the hot water to lower the overall average temperature of the liquid water. Did you ever put an ice cube into hot tea to cool the tea just enough so it wouldn't burn your lips? When all of the ice has melted, a new but lower overall temperature is established uniformly for all the liquid water (i.e., presumably all temperature gradients have in time worked their way out of the system).

Under these conditions, we say that the hot water-ice cube system has relaxed back to a state of *thermal equilibrium*. This relaxation process stops when there are no longer any thermal gradients to drive the process of change. The developers of thermodynamics (Kelvin, Celsius, Carnot) realized that since the entire universe is a closed system and its entropy continues to always increase without limit (until all of its temperature gradients have disappeared!), then there must come a time in the distant future when our entire universe will reach thermal equilibrium.

This final chapter in our universe's evolutionary saga is known to science as heat death. The concept of our whole universe plunging inevitably into heat death has long haunted the psyches of many scientists in nightmarish ways. (However, theologians would say that an end to the universe only indicates that our material universe is *not* eternal, unlike God.) However, an important conclusion to come out of this line of reasoning is that the increasing entropy (by way of the second law) seems to be closely associated with the passage of time. Because total entropy is always increasing, time likewise must always be *increasing* (move forward and never backward). Time travel to the past is not allowed in thermodynamics! This ever-increasing (and never going back) flow of time is known to science as the arrow of time. Thought, however, can travel back in time, as all historians know.

Let us examine more closely how time and entropy relate. All closed systems must increase their entropy until thermal equilibrium is reached. However, a closed system can never be made to run backward; that is, it cannot be made to return to its initial conditions. Systems, like people, can never go home again. This is because the process of entropy itself wipes out the system's memory of the past by increasing its disorder. Since there remains no record of where the system has been to serve as a map for the return trip, any return trip becomes impossible.

Oddly, there will always remain plenty of energy within the system (remember the first law's admonition that energy cannot be created or destroyed) to fuel future activity. However, without a sense of its own history, and without the presence of thermal gradients (after the system has returned to thermal equilibrium) to drive new actions, it becomes impossible for a thermodynamic system to return to its past. When you are a part of our material universe, you simply can't go back; there is no returning to the past.

It is very interesting to note that unlike the unidirectional time arrow of thermodynamics, the Schrödinger wave equation[55] of quantum mechanics (as well as Newton's mechanics) is completely reversible in time. Quantum mechanical systems (prior to measurement) are completely reversible in time; and time in the quantum world freely flows either backward or forward with equal ease. Obviously, there appears to be a fundamental difference between what quantum mechanics calls time and what thermodynamics calls time. Remember that in the quantum world, prior to measurement, you can always go back. It is in the measurement process itself that literally throws a quantum mechanical system into the *irreversible* world of thermodynamics, where it also can never go home.

Statistical Mechanics

By the 1880s, the field of thermodynamics had moved on from being the exclusive territory of steam engine designers to becoming a part of the mainstream fields of physics and chemistry. Physicists like Ludwig Boltzmann in Austria and James Clerk Maxwell in Scotland developed a keen interest in thermodynamics. However, these physicists took

---

[55]   In quantum mechanics, the Schrödinger wave equation enables the calculation of how a system's wave function evolves in both space and time. In terms of equations, the Schrödinger wave equation is expressed as: $H\psi = i\ h\ (d\psi/dt)$ where H is the Hamiltonian operator, ψ is the wave function, i is the square root of -1.0, and h is Planck's constant (6.626E-34 joules-seconds).

thermodynamics in a different direction than the steam engine designers did.

As we have already mentioned, Boltzmann was struggling with the paradox of the arrow of time, which seemed to completely contradict the time reversibility of Newton's mechanics. Unlike the macroscopic phenomenological approach taken by the steam engineers, the physicists took a microscopic approach to the problem, making use of the still-controversial atomic theory. This new approach to thermodynamics came to be known as statistical mechanics because the physicists used statistical methods that were applied to large populations of atoms and molecules to calculate the resulting system's macroscopic thermodynamic properties such as temperature and entropy.

For the physicists, thermodynamic systems were the large populations of atoms or molecules that are contained in gases and liquids. The first application of statistical mechanics was to gases, but like most important new theoretical pictures in science, statistical mechanics has been applied to many other gas-like systems, such as the electron cloud that travels through a metal or a semiconductor during the operation of the transistors and diodes that make up integrated circuits. Hard as it is to believe today, Boltzmann's initial work drew a lot of criticism, not because of his use of statistical methods but because his approach to thermodynamics made use of the then-controversial theory of atoms! The reality of atoms had not yet been fully accepted by science, and this aspect of Boltzmann's work raised a lot of eyebrows! In the 1880s the so-called "atomic hypothesis" was still very controversial.

There was a popular philosophy of science at this time that is now called *rational positivism*. This philosophy of science was most closely associated with another Austrian physicist named Ernst Mach.[56] Mach felt atomic theory was wholly unnecessary for the advancement of science. From Mach's point of view, atoms were at best a useless distraction; at worst they were highly misleading and an unnecessary complication. For

---

[56] Ernst Mach (1838–1916) was an Austrian physicist and philosopher. He is best known for a philosophy of science called rational positivism. Rational positivists were very opposed to atomic theory as unnecessary for the advancement of science. Nevertheless, rational positivism had a certain degree of influence on both Niels Bohr and Weiner Heisenberg and therefore on the development of quantum mechanics from the Copenhagen perspective.

many years, Boltzmann was on the receiving end of constant criticism from the rational positivists. In the end this storm of constant criticism drove him around the bend, and he committed suicide while still a young man. Fortunately for the future of science, Boltzmann's death came after he had completed the majority of his work on statistical mechanics.

Let us now consider some of the finer points in Boltzmann's work. In statistical mechanics, the mathematics of statistics is applied to the atoms and molecules of a gas to predict the gas's macroscopic properties (like temperature, entropy, energy, etc.). By calculating certain key statistical averages, the macroscopic properties of the gas (such as temperature) become associated with the distribution of its particle's energies. In statistical mechanics, a high temperature indicates an energy distribution tilted toward high energies, and a low temperature indicates an energy distribution tilted toward low energies. Zero temperature, if it could be achieved, would predict a population where all particle movement had simply stopped.

Unlike the probabilities associated with quantum mechanical wave functions (truly reflecting a fundamental kind of unknowing), the probabilities of statistical mechanics deal with a more everyday kind of unknowing that is related to our practical inability to keep track of such a large number of individual atoms and molecules comprising this vast population. The billions of atoms and/or molecules that make up even a very small amount of gas are each, at least in principle, completely describable by the mathematical techniques of the classical physics used by Newton. However, in practice such an immense computation is beyond the abilities of even the most powerful supercomputer.

Perhaps the crowning achievement of statistical mechanics was Boltzmann's approach to the calculation of entropy. Boltzmann's technique for calculating the entropy of a gas turned out to be very simple mathematically, but conceptually the Boltzmann entropy calculation followed a long tradition of stretching our minds to understand the concept of entropy.

Boltzmann imagined the situation in which a cloud of gas particles were situated so that they were free to arrange themselves in any number of different ways. You might think of these particles as a collection of stones that could be arranged on the ground in a great variety of different patterns (try to imagine what a vast number of different ways all of the grains of sand on a beach could be arranged). The sorting of the gas particles making up such a statistical arrangement was accomplished not only in terms of each particle's many possible positions but also in terms of the momentum associated with each individual particle.

To accomplish his purpose, Boltzmann created an abstract six-dimensional mathematical space called "phase space." Six-dimensional phase space is comprised of three spatial (position) dimensions and three momentum dimensions (each heading in the same direction as its equivalent spatial dimension). It turns out that phase space is an ideal mathematical device for calculating the arrangements of particles within a gas. Boltzmann called one of the many possible arrangements of the gas particle population (within phase space) a "complexion." Complexion was simply Boltzmann's term for one of the possible arrangements of a large population of gas molecules.

Based on his extremely abstract and inventive approach to calculating entropy, Boltzmann was able to determine the entropy of any gas to be simply a constant (called Boltzmann's constant, which has a value of 1.38 E-23 joules/degrees Kelvin) times the logarithm of the number of complexions (i.e., arrangements) that can be assumed by the gas. Boltzmann's mathematical definition of entropy succeeded in establishing an intimate relationship between entropy and complexity. From Boltzmann's point of view, the more different ways a system of particles could be rearranged, the greater would be the system's complexity and therefore its entropy.

By Boltzmann's definition, entropy becomes a measure of complexity and complication. In this regard, the second law of thermodynamics, as interpreted by Boltzmann, tells us that the complexity of the universe must always be increasing! To anyone returning to the office on Monday morning after a long weekend away, Boltzmann's concept is no surprise. This new understanding does shed some light on what might be happening as the universe proceeds forward toward its ultimate fate of heat death. The real meaning of heat death concerns how our universe is constantly increasing in complexity as time unfolds.

Increasing complexity implies a universe that is getting progressively finer and finer-grained texturally and is distributing itself in an ever-increasingly more uniform way. Vast seas of totally uniform chaos are replacing islands of order. These thermodynamic realities exist in almost complete contrast to the harmonious, almost-musical relationships that are present within the coherent quantum mechanical systems (whose elements, like the instruments in an orchestra, seem to obey a common conductor). In complete contrast to the quantum mechanical systems of the subatomic world, the thermodynamic universe is constantly being driven by the arrow of time into the inevitable direction of total chaos.

Maxwell's Demon

During the second half of the nineteenth century, not all scientists were completely convinced of the truth of the second law of thermodynamics. James Clerk Maxwell in Scotland proposed a Gedanken experiment that questioned the universal validity of the second law. Maxwell himself was not personally convinced of the universal validity of the second law, but he wasn't really sure one way or another. This experiment was designed to raise the issue of the second law in front of the entire scientific community.

In the grand tradition that was to be followed by Einstein and many other scientists, Maxwell proposed a thought experiment as a way of exploring possible weaknesses in what had come to be regarded as a universal law. Maxwell's Gedanken experiment has come to be called the Maxwell's demon experiment, and it has succeeded in torturing several generations of scientists until the demon was finally exorcised by Leo Szilard[57] in the late 1930s.

In the Maxwell's demon thought experiment, a devilish creature with uncanny perceptive abilities plays a key role in the experiment's execution. The experiment is arranged as follows. Suppose a chamber containing an ideal gas is divided into two subchambers. The subchambers are connected by a trap door, which can be opened or closed from the outside. The demon, having an excellent view of both chambers, performs the task of opening or shutting the trap door.

Let us suppose the left-hand subchamber contains a hot gas, which is composed of a population of highly energetic gas molecules, and the right-hand subchamber contains a cold gas that is composed of a population of slow-moving, less-energetic gas molecules. The demon, whose avowed intent is to defeat the second law, will attempt (in violation

---

[57] Leo Szilard (1898–1964), a Hungarian and American physicist, was best known for his work on thermodynamics and nuclear physics. Together with Enrico Fermi, Szilard constructed the world's first fission nuclear reactor in 1942.

of the second law) to make heat energy flow from the cold subchamber to the hot sub chamber.

The demon plans to accomplish this feat by means of opening the trap door whenever there is an opportunity to allow a rare highly energetic gas molecule to pass from the cold subchamber on the right into the hot subchamber on the left. Likewise the demon will open the trap door to allow a rare low-energy molecule from the hot sub-chamber to pass into the cold subchamber. In this way, the demon is determined to heat up the hot subchamber at the expense of cooling down the cold subchamber still further, thereby defeating the second law of thermodynamics! If the demon succeeds in this enterprise, the second law will have been overturned because heat energy will have been forced, by the demon, to flow from the cold subchamber into the hot subchamber in violation of the second law!

Very few scientists in Maxwell's day believed the second law was in any real danger of being overturned by the Maxwell demon experiment. Most scientists of the day, perhaps including Maxwell himself, viewed the Maxwell's demon experiment as a kind of scientific practical joke, which it no doubt was. However, no one could come up with a credible way to refute the Maxwell's demon experiment. It is almost embarrassing how long it took to find a solution to the paradox of Maxwell's demon.

The Demon Is Exorcised

As unbelievable as it sounds today, this challenge to the second law of thermodynamics stood unanswered until the 1930s, when a Hungarian physicist named Leo Szilard cleared up the confusion and exorcised the demon once and for all.

Szilard recognized that the demon would be unable to see individual gas molecules without the benefit of some kind of highly specialized flashlight. It turns out that both the hot and the cold subchambers of the experiment were literally being flooded with infrared light associated with the thermal agitation of the heated gas molecules. Under these demanding conditions, the demon's flashlight would be required to provide a light beam of much higher energy than the gas's own infrared radiation (emanating from the gas molecules). Only when armed with his high-energy flashlight could the demon gain a view of the gas particles that would allow him to distinguish an individual gas molecule against the gas's infrared background radiation.

This situation would surely require the demon to possess a flashlight producing light in the visible portion of the electromagnetic spectrum (perhaps like ourselves, the demon's eyes respond most strongly to visible light). In keeping with Einstein's 1905 analysis of the photoelectric effect, each photon (i.e., the quantized light particles emanating from the demon's flashlight) would contain an energy equal to Planck's constant ($h=3.43 \times 10^{-38}$ joules-seconds) times the frequency of the light (or $E=hF$ in equation form). In order for the demon to "see" individual gas particles, this photon energy would have to be significantly higher (or at the very least equal to) the average energy of the gas molecules (the energy that is associated with the gas's infrared background radiation). This energy is equal to Boltzmann's constant ($k=1.38 \times 10^{-23}$ joules per degree Kelvin) times the gas's temperature ($E=kT$ in equation form). Therefore, our hopefully well-informed demon must expend (via his flashlight) a photon energy of at least $E=hF$, which is greater than kT, for there to be a chance for the demon to locate individual gas molecules.

Here is how the demon and his criticism of the second law are defeated. When the demon's flashlight shines into the cold subchamber, the flashlight's photon energy will more than replace the energy lost by the cold subchamber via the exiting energetic molecule. Therefore, rather than cooling the cold subchamber by the removing of an energetic molecule, the demon has actually succeeded in raising the total energy of the cold subchamber, which means that in reality the heat energy is flowing from the hot subchamber to the cold subchamber, in keeping with the second law.

This conclusion, which is Szilard's principal insight, preserves the validity of the second law, and as an added bonus, it proves to all who are interested that one bit of information is always associated with an energy of at least kT joules. Remember, the demon had to gather at least one bit of information about the gas molecule before he could decide whether to open or shut the trap door. This universal conclusion has been extended to include many areas of communication and information theory. In all areas of communication theory, it is universally required (by the second law of thermodynamics, no less) that for any and all communications systems, at least kT joules of energy must be associated with each encoded bit of information. This is a very important conclusion that will have profound implications for our search for a unified truth.

The Great Paradox

Here is the origin of the great paradox: for God to communicate with our material universe, we must assume that God's communications also obey these same restrictions to remain completely consistent with the laws of physics. But here is the rub: if God introduces one bit of information into our universe, God must also introduce at least kT joules of energy into our universe, since this energy is necessary for sustaining the bit of information. In the process, this introduction of energy into our universe by God automatically violates the first law of thermodynamics (to wit: energy can neither be created nor destroyed within our universe). Now we have reached the great paradox that God must, it seems, violate his own first law of thermodynamics to communicate with our universe! How can this be? Remember, if God could not communicate with our universe, then we would have no knowledge of God's reality and presence. Without this knowledge, we would all be set adrift in a sea of ignorance!

God Communications

The great paradox leads us straight into a godly interpretation of the laws of science. Whenever God acts in our universe, for any of us to be aware of these actions, God's actions must appear somewhere in the form of a body of information. For example, the tablets Moses received on Mt. Sinai contained writing (i.e., information). However, every bit of information in our physical universe has some amount of energy associated with it. So if God is to act by communicating with our universe, then God must cause some measurable amount of energy to spontaneously appear somewhere out of nowhere in our universe. But this requirement is in direct conflict with the requirements of the first law of thermodynamics, which tells us that energy can neither be created nor destroyed within our universe.

So how does God do it? How does God act in our universe without violating the first law of thermodynamics? Without the appearance of some form of information, we would never have known that God had acted.

(Moses would have been clueless about God's intentions for him had the burning bush neither burned nor spoken or if nothing had been written on the tablets on Mt. Sinai!) But it is the very appearance of information conveying God's actions and plan that requires the spontaneous appearance of a measurable amount of energy in direct violation of the first Law. We must conclude, therefore, based on thermodynamics alone, that God's information-generating actions will violate the first law of thermodynamics. This calls all of science into question.

If you overturn such a fundamental pillar of science as the first law, you overturn all of science. How can we break this paradox and still make a case for the knowledge of God acting within our universe that is consistent with both our scientific understanding of the laws governing the physical universe and our religious understanding of the necessity for communications from and companionship with God?

The argument for a minimum amount of energy associated with a bit of information is based on the second law of thermodynamics. Albert Einstein once remarked that even if all of the other theories of science were to be disproved or replaced, the second law would still remain. The basis for the second law is rooted solely in the separation and un-connectedness of the particles that make up our material universe. Everyday, macroscopic examples of the second law abound. For example, when a book is printed, the press expends energy during the printing process, fixing an amount of information onto the book's pages. The book is read and reread over a period of years and eventually wears out or is damaged. In the end, its information (order) is irreversibly converted into entropy (disorder) as the book is destroyed.

But how is God's message communicated to the people of our small planet? How can we come to understand this greatest of mysteries? The answer to this question must unlock an entirely new way of looking at the laws of science and how the truth of science and the truth of religion, rather than contradicting each other, affirm each other's truthfulness.

CHAPTER 4

RESOLVING THE GREAT PARADOX

Next we will explore ways to reconcile God's communications into our natural world with the restrictions placed on these communications by the scientific laws of thermodynamics and information theory. For religious and scientific truths to be truly unified, it is absolutely essential that God's ability to communicate with our universe not be restricted in any way. On the other hand, to preserve the truth of science, all of the laws of thermodynamics, information theory, and quantum mechanics must be simultaneously upheld. Like the cloud patterns of the dancing Einstein changing into the soaring eagle, we need to conceive of ways to visualize how God's unrestricted communications with our natural world can be made completely consistent with all of the most fundamental laws of science. As we are about to see, the answer to this most profound of questions lies deep within the mysteries of quantum mechanics.

Information and the Natural World

Our particular location within our natural world is a place dominated by space, time, mass, energy, and information. Information is everywhere. It is in our books, in our newspapers, in our computers, in our cell phones, in our maps, in our televisions, in our radios, in our cars, and in our airplanes. It is also in the natural world that is all about us in the form of plants and animals, mountains and deserts, oceans and glaciers, stars, planets, and galaxies, and nearest of all to us, our own bodies. The list is endless because information is literally everywhere!

Information is not a thing in itself (or is it?). As we saw in the last chapter, the resolution of the Maxwell demon's paradox proves that it takes a small amount of energy to encode one bit of information. But information is the blueprint from which each and every thing is constructed. Information conveys an idea from one individual to another individual. Information is a lot like the Greek philosopher Plato's[58] concept of the ideal, which is a kind of descriptive perfection that can never be fully realized; it can only be approached in the real world.

Information allows us to share our ideas with our fellows. When God communicates with human beings, God's communications must be the result of some kind of transfer of information that originates with God but is addressed to all of us and/or just one of us. But what exactly is information? How can God impart information, since God (who exists outside our natural world or else would be trapped in it) does not violate any of the laws of science (which are, after all, God's own laws for our natural world)?

Information and Science

Let's step back for a minute and look at what science can tell us about information. Up until the middle of the twentieth century, science had very little to say about information. Then two important scientific events took place in the years just before, during, and just after World War II. The first event was the development of a mathematical theory of information by Claude Shannon,[59] a Bell Telephone Company communications engineer. Shannon was a highly talented mathematician, and he soon realized that information was composed of, in its most elemental form, a series of choices between two or more alternatives. Shannon reasoned that by making a choice between alternatives, information was reducing

---

58 Plato (423–347 BCE) was an ancient Greek philosopher. He is best known for his philosophy of the ideal that bears his name (Platonic).

59 Claude Shannon (1916–2001) was an American communications engineer and mathematician. Shannon is best known for his work on the mathematical theory of information, which has in fact given birth to the full flowering of the information age.

uncertainty. Uncertainty increases with the number of choices that need to be considered before making a decision.

When there are just two choices, there is not that much uncertainty (a heads or tails choice is always a fifty-fifty proposition). However, when there are one hundred possible choices, the uncertainty associated with such a choice increases dramatically. A choice between just two alternatives is what Shannon called a bit of information (bit is an abbreviation for binary information). Using the mathematics of logarithms, Shannon, in a fantastic leap of intuition, calculated that the amount of information contained in any set of choices as the logarithm (to the base 2) of the number of choices that need to be considered when answering the question.

For instance, a choice between two alternatives (LOG2=1) contains 1 bit of information, and the choice between four alternatives (LOG4=2) contains 2 bits of information. Logically it follows that a single alternative, which is to say no choice at all, contains (LOG1=0) 0 bits of information. This conclusion is at the very heart of information and communications theory. Depending on how you look at it, the presence of information always requires one to make a choice or answer a question, which is pretty much saying the same thing.

When information is created, something new must always appear. If the outcomes were already known, there would be no new information. Information is like news: if everyone knows about it, it is no longer news. Information is not contained in the alternatives themselves but in the making of a choice between two or more alternatives. Some examples of information are the choice between a 1-volt signal and a 0-volt signal in an electrical circuit or the contrast between black ink and white paper on the printed page you are reading right now or the choice between a lit phosphorescent pixel and an unlit phosphorescent pixel on a computer monitor's screen.

At the urging of his old friend Norbert Wiener,[60] Shannon named his mathematical measure of information after the thermodynamic (second law) measure of disorder: entropy. However, it turned out that Shannon's choice of entropy as the measure of information was more than just a random choice.

---

[60] Norbert Wiener (1894–1964) was a Swedish-American physicist best known for his invention of cybernetics.

First of all, in the statistical mechanics of a many-bodied system (e.g., a gas containing billions of molecules), the entropy of the gas would be equal to a constant (called Boltzmann's constant, which is equal to 1.38 x 10^{-23} joules per degree Kelvin) times the logarithm of the number of possible arrangements of this system (a very large number indeed!). In an analogous way, Shannon's information measure is a constant, 1.0, times the logarithm of the number of possible choices required to answer a question.

We can see from the start that a fundamental connection exists between thermodynamics and information. In fact, thermodynamic entropy can be viewed as the measure of how much information is lost when a message becomes jumbled, in effect destroying the useful pattern of its information. However, its mathematical information content is preserved.

For example, put a page of text into a paper shredder. The useful information on the page is lost during the shredding process, but the amount of mathematical information on the page is preserved because a paper shredder turns the useful message on a printed page into disordered information—or entropy—as all those little pieces of paper, like a vast jigsaw puzzle, are reduced to a jumbled state that can never be reassembled to their original form. Another way to visualize this process is to remember the nursery rhyme describing what happened to Humpty Dumpty. We might choose to think of thermodynamic entropy as lost (or hidden from view) information.

Of course, under this definition of information, what constitutes a message can be defined very generally indeed (for instance, consider how the arrangements of different colored molecules within a liquid might be considered to be a message). However, it is important to remember that although information has a precise mathematical description, like beauty, the real message must always be in the eye (and heart) of the beholder.

The second significant scientific finding concerning information happened just before the start of World War II when (as we discussed in the previous chapter) the Hungarian physicist Leo Szilard solved a nearly seventy-year-old physics paradox called the Maxwell's demon paradox. By solving the Maxwell's demon paradox, Szilard was able to show, in a way that is just as fundamental as the second law of thermodynamics (which it upholds) that some minimum amount of energy must always be associated with each and every bit of information.

How do these mathematical and thermodynamic restrictions on information affect God's ability to communicate with our material universe and with ourselves? First of all, assuming God wishes to avoid violating the first law of thermodynamics (conservation of energy), God must refrain from spontaneously introducing into our material universe even the smallest amount of energy associated with each new bit of information. To answer a question by making a choice, God must act in such a way that each bit of godly information is represented by a choice between some physically distinguishable configurations that exist within our natural world. However, based on Szilard's solution to the Maxwell's demon paradox, each bit of godly information must in some way be associated with an energy increase equal to Boltzmann's constant times the ambient temperature of the environment into which the information is being delivered.

The only conceivable way for all of these conditions to simultaneously be met is to rely on quantum mechanical uncertainty. Quantum uncertainty can make it possible for the required amount of energy associated with a bit of information to arise naturally (like virtual particles[61]) without the need for this energy to be injected into the universe unnaturally by God. The injection of energy will inevitably result in breaking the laws of science and thereby destroying science's foundation and truth. That is the same thing as saying the laws of science work some of the time but not all of the time! The situation can only be made workable (and the paradox resolved) to the satisfaction of all parties (both religious and scientific) if godly information is passed into our material universe as the direct result of a scientifically a-causal single quantum mechanical event whose final outcome must remain unknown and unknowable until a final measurement is made at the exact moment of wave function collapse! In the final analysis, it is Heisenberg's uncertainty principle that saves the day.

---

[61] Virtual particles are a quantum mechanical phenomenon whereby pairs of particles spontaneously come briefly into and out of existence

A Gedanken Experiment in Godly Communications

Let us now consider an example of how information might be transferred from God to our natural world by a quantum mechanical process without violating any of the laws of science. Our example (which admittedly is a highly conceptualized, oversimplified demonstration of our point) is a quantum mechanical measurement system consisting of a source of electrons, a metal plate containing two very small holes (one beside the other), and a photosensitive plate. Refer to the illustration on page 64.

The diameter and spacing of the holes are comparable in size to the wavelength of the electrons. The sensitive photographic plate is located behind the metal plate and serves the purpose of recording electron impacts. We might think of this measurement system as two tiny pinhole cameras, with each camera designed to record the impact of electrons instead of recording visible light images. By thinking quantum mechanically, we will associate a superposition of two wave states with each electron. One wave-state will go through the left hole in the plate, and the second wave-state will go through the right hole in the plate.

It is an experimentally established fact—demonstrating what has come to be known as quantum weirdness—that even when such an experiment is conducted using only a single electron, the measurement process must be thought of as the coherent interference of the superposition of the left hole wave-state and the right hole wave-state passing through the plate simultaneously. Yes, in quantum mechanics an electron truly can be in two places at once. Once the two wave superpositions have passed through their respective holes, they are free to coherently interfere (interact) with each other before reaching the photographic plate, where a measurement will take place in the form of a smudge on the photographic plate.

Quantum mechanics tells us that during the course of this experiment, the electron quite literally interacts with itself! The electron, in the form of a wave superposition of the left hole wave and the right hole wave will interfere with itself before the measurement is completed. This experiment is a good example of how quantum mechanics defies our everyday sense of intuition about the natural world. With great wisdom, Albert Einstein referred to the quantum wave function as the "ghost function."

FIGURE 3:

Gedanken Experiment in Measuring Divine Action

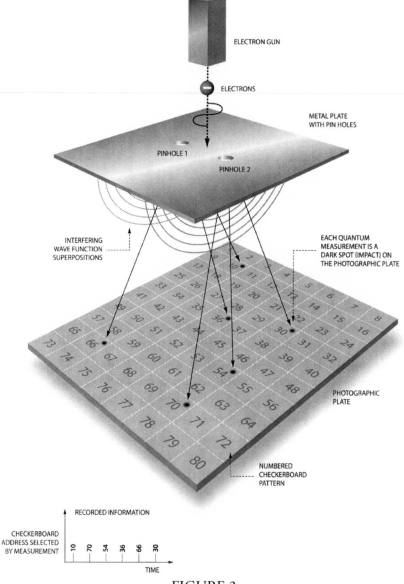

FIGURE 3:
ILLUSTRATION OF A GEDANKEN EXPERIMENT
IN GODLY COMMUNICATIONS

At the photographic measurement plate, where results of the experiment are recorded, the self-interference of the two wave-function superposition will take on a pattern of constructive and destructive interference "rings" radiating out from the two holes in elliptic patterns. The regions of constructive interference correspond to those areas where there is a high probability of an electron impact. The regions of destructive interference correspond to areas where there is a low probability of impact. In an actual experiment of this type (this experiment is in fact very similar to the famous double-slit experiment discussed in chapter 2), the regions that have the highest probability of being impacted are the regions of the photographic plate lying directly behind a point midway between the two holes.

Of course, our everyday intuition tells us that this point is the least likely place for a particle-like electron to impact the photographic plate. (Because of its particle nature, the electron is expected to come through one hole or the other but not straight through the wall of the metal plate!) However, the fact is that because of the probabilistic nature of the quantum wave function, in the end only God knows where the electron will actually impact the photographic plate.

All we can really say is, based on the laws of quantum mechanics, some kind of topographical map demonstrating probability of impact will be created as smudges on the photographic plate as more and more electrons are passed through the experiment. Such a map would show the locations where an impact is very likely and also the locations where an impact is highly unlikely. However, the exact location of the point of impact for an individual electron cannot be determined from the laws of quantum mechanics (or by any other law of science, for that matter) as a direct result of the probabilistic nature of quantum mechanics.

In this experiment, the best that science can ever hope to say about the location of an individual electron's point of impact is that only God knows where the impact will occur. Therefore, we can conclude that ultimately it is God's choice alone that determines where an individual impact will occur. Here is the key idea to consider relative to godly communications: it is by making of these quantum mechanical choices that God creates a mechanism (for lack of a better term) to transmit his message into our natural universe. Conceivably, if God so chooses, he could encode a message of his own choosing by using this experimental arrangement to make the impact sites serve as a series of dots and dashes that could transmit his message as a kind of visual Morse code.

A more sophisticated version of this experiment might involve the creation of a two-dimensional pixel checkerboard pattern drawn on the photographic plate in which each "pixel square" represents a choice. If there are P pixels on the photographic plate, the amount of information associated with each electronic impact becomes (according to Shannon's mathematics of information) LOG(P).

By making these choices, God in no way violates the thermodynamic law of conservation of energy because the exact amount of energy of each electron remains unknown until its wave function collapses. Until the measurement is actually made (in this case by recording a black spot on the photographic plate at the exact point of impact), the electron exists in a state of unreality where the electron's behavior runs counter to our normal, everyday understanding of matter and energy. It is only at the point when a measurement occurs (i.e., an electron impacting the photographic plate) that the electron resumes its particle nature and joins the everyday world of space, time, matter, energy, and information. It is determining the exact location and timing of each electron's impact on the photographic plate that enables God to write a message into our natural world.

We must bear in mind that a single quantum event in an experiment of this kind is only one of billions and billions of similar subatomic events that are taking place all around us and inside of us at every moment of the day. For example, there are about as many synapses within the human brain as there are galaxies in our entire universe (in excess of 100 billion). If each synapse in any individual's brain were to be influenced simultaneously by God, then a very great quantity of information could potentially be imparted by God to an individual. Each one of these synaptic events could potentially represent one bit of information that God is communicating to us and to our world.

God's work in the natural world often seems to adopt an approach that makes it necessary to use a high degree of redundancy for the purpose of getting the job done. For instance, consider that during mating, it takes only a single sperm to fertilize one egg at conception. However, millions of sperm are released by the male during mating, ensuring that at least one sperm will succeed in its mission of creating new life (even if hundreds of thousands of other sperm fail in their mission to fertilize an egg).

It is much the same with information. By reserving the right to exercise quantum mechanical choices during subatomic events, God has reserved vast opportunities for communications with our world and with each of us personally. Quite literally, the universe is alive with the word of God! When and where God chooses to exercise these communications is up to God, with our part being that of a good listener. God's ability to influence the flow of information at many times and places simultaneously is surely the source of God's omnipotence. Traditionally, religion has regarded God's omnipotence as being more in line with throwing lightning bolts and spinning nebulas. However, we believe that it is God's constant influence at the very core of our material universe that is most truly the source of our awareness of God's omnipotence.

Quantum Synchronicity

A noncausal event theory of synchronicity was developed jointly by Carl Jung,[62] who was a Swiss psychologist, and Wolfgang Pauli, a quantum physicist. Jung and Pauli may seem like odd bedfellows, but their collaboration brought to light a very powerful way of looking at the human experience and the workings of the human mind.

Their collaboration began when Pauli landed on Jung's couch because of the many difficulties that were going on in Pauli's private life. Pauli had been a leading light in the development of quantum mechanics and atomic physics (the Pauli Exclusion Principle is one example), but on a personal level, he was something of a prickly pear. He could be very critical of his friends and enemies alike, and as a result, his friends (and his enemies) treated him as a laughingstock.

Pauli had a reputation for being a poor experimentalist. In fact, it was rumored that Pauli had such a black finger in the lab that experiments would fail just because he was standing in the same lab. His reputation grew so extreme that it became a standing joke that every time an important experiment failed, it was later found that Pauli had at that moment been on a train less than ten miles from the site of the experiment.

---

[62]   Carl Jung (1876–1961) was a Swiss psychiatrist and one of the founders of analytic psychology.

At some time in his life, Pauli impulsively married a dancer with whom he had absolutely nothing in common. The marriage lasted only a few weeks, and when it was over, Pauli was forced to suffer the humiliation of admitting to his colleagues that he had made a serious mistake.

Although Pauli's family background was Jewish, his parents had converted to Catholicism many years before. Pauli was raised a Catholic. It was during this period that Pauli become very disenchanted and left the church. He began to feel utterly lost and ultimately turned to Jung for advice and counsel. We don't know how successful Jung's therapy was, but we do know that the two men cooked up the idea of synchronicity during Pauli's therapy. It is very ironic that one of the cofounders of quantum mechanics also became the cofounder of this important new psychological theory. Perhaps the Jung-Pauli collaboration hints that quantum mechanics has a wide area of application, well beyond atomic physics.

As Jung and Pauli conceived it, synchronicity is an a-causal experience in which an individual encounters the coincidence of coming upon the same word or topic within a short period of time from a number of different and totally unrelated sources. For instance, a man may read an article about fish in the morning paper, find that his lunch is a fish fillet, and get a call from an old friend that evening inviting him to go on a weeklong fishing trip.

Jung's personal experience of synchronicity is related in the story of a woman who was Jung's patient. The woman, who had taken a very analytic approach to life, wanted to explore other, more emotional and more spiritual sides of herself. Jung recommended that she record her dreams for a few weeks, and the two of them would discuss her most important dreams during her appointments. One of these dreams involved finding a beautiful Egyptian scarab bracelet. As she related her dream to Jung, he heard scratching at the window behind him. Jung got up to open the window, and in flew a large scarab beetle! The beetle landed in Jung's hand, and he walked over to the couch where his patient was still describing her dream and said, "Mrs. X, here is your scarab."[63]

We can think of God's intervention in quantum measurements as an abundance of a-causal events. God is not a thing in the ordinary sense of the word, and so God should not be thought of as a cause. What God is

---

[63] Joseph Campbell, *The Portable Jung* (New York; Penguin Books, 1971), 511–12.

about in quantum measurements is an enabling source that brings about information. Since God is not a part of our material universe ("I am in this world but not of this world" [John 1:10-14, 8:23]), God does not add to or subtract from our material universe in any way. Rather, God influences the exact composition of the information that was already in the process of formation.

God's ability to affect the composition of information during quantum measurements may be synchronized over space and time with other quantum measurements. (Recall, there are as many as one hundred billion synapses in the human brain.) It is the net effect of this array of quantum measurements being synchronized that is experienced by individual humans.

We might compare this process to a flock of birds or a school of fish. Although there are no discernable communications between the birds of the flock or the fish of the school, the individuals behave in such a way that each flock and school as a body appears to take on a life and purpose of its own. Flocks and schools twist and turn as they please, almost as if they had become a single individual with a common will that overrides the will of the individuals within it.

In the same way, God's information, as it comes into being during a large number of synchronized quantum measurements, will appear to have an existence of its own when experienced by individual humans (who are watchfully sensitive to this sort of thing in their own practices of prayer and meditation). This godly information will inform the individual of God's presence and plan by the observation of both internal and external experiences. We may think of synchronized godly information as a point cloud of bits of information in space-time evolving dynamically in a direct analogy with flocks of birds and schools of fish. However, as in all events that are linked to God's communications, timing is everything when discerning the meaning of these communications for a particular individual.

From Determinism to Faith: A Pathway Leading out of the Shackles of Certainty to the Freedom of Revelation

In about 1630, the French philosopher Rene Descartes[64] found himself fighting in rural Germany as a soldier in the Thirty Years' War. Descartes had already reached his famous conclusion about human existence that simply says, "I reason; therefore I am." According to the story (which may be little more than a philosophical myth), one day while in a small German town, Descartes spent an entire night in a ceramic furnace (it may have been a kind of sauna or steam bath) and was visited by a series of wild dreams and hallucinations. His most vivid dream was for the future of the field of physics. In Descartes's dream, the field of physics became highly dependent on mathematical methods. Descartes's dream heralded a future where the predictive power of mathematics would replace observation and measurements as the most powerful tools for advancing the field of physics.

As a kind of validation of Descartes's dream, during the next two hundred years France and Germany were destined to produce many individuals who are now considered to be among the world's greatest mathematicians. This was the age of Fourier and Laplace, of Legrange[65] and Hermite,[66] of Foucault[67] and Euler,[68] of Bessel[69] and Leibnitz,[70] and of Gauss.[71]

Growth in the developing fields of physics and chemistry was in no small measure the direct result of advances in mathematics, which was necessary to provide the physical scientists with the necessary mathematical tools to fuel growth. This was also a period when many scientists envisioned what has come to be called a clockwork universe.

---

[64] Rene Descartes (1596–1650) was the French philosopher and scientist who famously said, "I reason; therefore I am." In science Descartes is best known for urging the field of physics to become more reliant on mathematical techniques.

[65] Joseph Louis Legrange (1736–1813) was a French mathematician..

[66] Charles Hermite (1822–1901) was a French mathematician.

[67] Michel Foucault (1926–1984) was a French mathematician

[68] Leonhard Euler (1707–1783) was a German mathematician.

[69] Friedrich Bessel (1784–1846) was a German mathematician.

[70] Gottfried Wilhelm Leibnitz (1646–1716) was a German mathematician.

[71] Carl Friedrich Gauss (1777–1855) was a German mathematician.

From the clockwork universe point of view, mathematically precise laws and their resulting predictions became direct replacements for the whims of a Creator God. This was the age of growing materialism (i.e., atheism). The mathematician Laplace is said to have famously answered Napoleon's question about the role of God in the physical universe with a terse, "I have no need for that hypothesis."

Science was becoming headstrong about its ability to understand all that is contained within our physical universe. For many scientists of this period (and of our own time, for that matter), God's only role lay in an initial act of creation. From the deistic point of view of these scientists (and many scientists today), after the act of creation, God stood aside and allowed the universe to evolve along the purely mechanical lines established by the mathematical laws of physics and chemistry.

This point of view has come to be called determinism. Determinism is what we call the ability of mathematical science to predict any and all future events based on the laws of science and a set of initial conditions. Perhaps the most powerful demonstration of determinism's power was the development of the laws of mechanics by Isaac Newton as set down in his book *Principia Mathematica*. Newton was able to demonstrate to nearly everyone's satisfaction that all matter in the universe, from apples to planets and stars, obey the same fundamental physical laws. Newton's mathematical predictions worked equally well for analyzing how a wagon is pulled by a team of horses or for analyzing how the planets of our solar system circle around the sun, each in its own unique orbit!

The educated world of his time stood in awe of Newton's accomplishments. Newton's tomb is located just to the left of the high altar in Westminster Abby in London. A visitor to Westminster Abby who is standing in the nave and looking forward toward the high altar will see Newton's tomb first. The grandeur of Newton's tomb is a powerful symbol of how the people of his time revered Newton as the greatest "saint" of science.

By the twentieth century, Albert Einstein had modified Newton's mechanics with the necessary corrections to enable mechanics to function accurately at the high relative velocities encountered in the vastness of intergalactic space. But even with all of Einstein's relativistic generalizations and corrections, Einstein's relativistic mechanics remained a deterministic science. Einstein (even though he was a lifelong free spirit in his personal life, vigorously objecting to any kind of social rigidity) held fast to his long-held belief that it was determinism that literally held the universe together. Einstein's deep and abiding objections to the new science of quantum mechanics were largely based on his strong belief in the primacy of determinism.

Quantum mechanics changed everything! With the development of quantum mechanics during the mid-twentieth century, determinism's stranglehold on the minds and hearts of scientists began to relax. Although the genesis of quantum mechanics lay in the desire of scientists to better understand the interactions of matter and energy at the subatomic level, the philosophical fallout from its development was destined to question all of the assumptions of determinism.

The uncertainty principle proposed by Werner Heisenberg and the probabilistic interpretation of the wave function proposed by Max Born and Louis de Broglie literally brought the age of determinism to an end. Heisenberg's brilliant insight was based on his recognition of the impossibility that any experiment could simultaneously measure the position and momentum of subatomic particles with total and complete precision. Heisenberg proved that the more accurately a particle's position is measured, the less accurately (and this is very fundamental) the particle's momentum can be measured. (For background, see the description of the single-slit experiment given in chapter 2). Alternatively, the more accurately one measures a particle's momentum, the less accurately one *can* (and this too is very fundamental) measure the particle's position.

The philosophical significance of quantum uncertainty is in its own unique ability to wipe out the possibility of determinism by eliminating any possibility of determining the initial conditions of a physical system at this most basic level. Since, based on quantum mechanics, all measurements must involve some element of uncertainty, it is impossible to know the exact starting point of any physical system, making it impossible for the clockwork evolution of determinism to begin.

The probabilistic interpretation of the quantum wave function is even harder to reconcile with determinism. Max Born and Louis de Broglie both realized that a quantum wave function could not be a measurable physical wave. (It is computed as a mathematically complex number, which means wave functions must be expressed in what mathematicians call *imaginary numbers* that comprise a whole new numbering system apart from the normal, real numbers of everyday life.) The waves in quantum mechanics demand a different understanding from other wave phenomena encountered in our physical world.

After much soul searching, the developers of quantum mechanics came to an understanding that a quantum wave function's amplitude is related to a probability that the quantum entity might be measured at some particular point in space-time. Remember that a quantum entity's wave nature is capable of extending over all of space and time, but its particle nature *must* be located at a particular point in space and time. What links these two natures is the wave function's ability to ascribe some probability to where the entity might be measured as a particle at a given point in space-time.

An example of such a quantum wave function might be a three-dimensional wave function that represents the orbits of atomic electrons circling around the nucleus of an atom. These wave functions appear to be clouds of fog enveloping the atom's nucleus. In fact, these wave functions are really clouds of probability that an electron might be measured at some specific point surrounding the nucleus. Quantum mechanics forces us to face up to the fact that at least on the subatomic level, the best knowledge we can ever hope to gain from our mathematical predictiveness is a calculation of the probability that something will be measured at some point in space-time. Therefore, we are forced to say, "So much for determinism!" But what will replace determinism? What is truth in the subatomic world?

Science has long tried hard to live with this probabilistic state of affairs that is forced upon it by quantum mechanics. But what really happens during the measurement of an individual quantum entity? It is exactly the same question as asking an election pollster to predict how John Smith, who lives at 1550 Easy Street in Omaha, Nebraska, will vote in an upcoming election. The best the poor pollster can say is that John Smith is 60 percent likely to vote for candidate A and 40 percent likely to vote for candidate B. There is no way for the pollster to say that John Smith is 100 percent likely to vote for one or the other candidate. Only John Smith knows that answer, and even he may not know for sure until Election Day.

Quantum entities work in exactly the same way. The outcome of an individual measurement is completely unpredictable. For this reason, science regards the outcome of each and every quantum measurement as being completely and totally random. In this case, random really has come to mean meaningless. A single quantum event is no more meaningful than a single throw of dice. But are such events truly random if God is involved?

Consider the alternative point of view. Consider that individual and coordinated quantum events are the way God enters our material universe to influence events in small but important ways by introducing bits of information at precise points in space-time. Think about the grandeur of it all! God's plan for introducing information into our material universe happens in a way of both great profundity and extreme subtlety!

By recognizing that God alone is the paramount controlling influence of the outcome of individual quantum events, we will come to recognize that these events serve to introduce God's message into the world in a way that does not violate conservation of energy (or any other law of science, for that matter). If God encodes his message in this particular way, it gets introduced into our universe at the most subtle of all levels.

The concept of godly control over quantum events gives us a new way of looking at what we mean when we say that God is omnipotent. All we human beings *can* ever know about individual quantum events is their probability of occurrence and never the certainty of their occurrence. There is always a small but finite chance that some very extreme outcomes might result. In such cases, the timing of God's actions is always the critical factor. What many people would consider to be a coincidence may be regarded by people of religious faith as a message from God.

We must remember that the total number of synapses in the human brain is on the order of the number of galaxies in the universe (over 100 billion) and the number of base pairs in the genetic code of the human genome is also on this order. When it comes to the human body, God has a very large canvas on which to paint his message. It is highly likely that all of the processes of change that occur in both the brain synapses and the genetic coding of our DNA are quantum mechanical in nature and are therefore subject to receiving ongoing, subtle messages from God.

Science regards all of these individual quantum events as entirely random and devoid of any meaning. However, it might just be possible for people to open their spiritual consciousness and come to regard these seemingly random events as the way in which God communicates with us. Try to imagine what kind of messages could result from the subtle effects of God's hand touching us simultaneously at an astronomically large number of points within our brains. Also consider how God could influence the evolution of our species and others as his hand touches the genetic coding of one generation after another. In both cases, God and living things truly become partners in the creative exercise of living and experiencing all that life has to offer.

By faith we can see a God whose subtle persistence is ever near to us, encouraging us onward and upward toward whatever is meant to be our place in God's plan. It is for us to choose whether we will see these subtle events as purely random or by choose, through faith, to see God's hand at work in these subtleties, nurturing us, training us, informing us, and changing us in ways most subtle and profound. By accepting the possibility of God's communication, we gain the richness of seeing beyond a purely material existence. We stand to gain a partnership of the material and the spiritual, leading us away from the controlling certainty (i.e., determinism) and toward the life-affirming freedom of a faith in God's presence that is always with us.

CHAPTER 5

GOD

This is a hard chapter to write. Who are we to attempt to describe the source of all of Creation? Who are we to ascribe to the maker and architect of the universe some kind of nature—a description that would be the sum of many parts? In light of what we have discussed in the previous chapter, we might regard God as the source—and the only source—of all the information flowing into our universe, including the information that becomes a part of us but originates inside of us.

During the Middle Ages, Jewish scribes would sit up all night praying for God's guidance before writing God's holy name during the next day's work. The Bible hints at the magnitude of what we are describing in the story of Moses's encounter with the burning bush. Moses asked the burning bush to tell him its name, and the bush simply said, "I am who I am" (Ex. 3:14).

A little later in the Bible, God, speaking through the prophet Isaiah (55:8) tells his people, "My thoughts are not your thoughts, and your ways are not my ways." God loves us, but God is definitely not one of us. It is a big mistake to think of God as being human, even a human with superhuman talents and potential. It is an even bigger mistake to apply human logic to the understanding of God ways.

Often there seem to be a strange kind mathematics at work when people attempt to describe God. Do most people conceive of God as having form or as being formless? C. S. Lewis[72] once remarked that all

---

[72] C. S. Lewis (1898–1963) was an English fantasy author and Christian theologian. Lewis is best known for his work *The Chronicles of Narnia*.

attempts on his part to conceive of a formless God led him only to the vision of an infinitely large bowl of tapioca!

It is so hard for us, with our human intellect and human experience, to conceive of God as formless. However, if God were to have form, then God must be composed of information. But this is out of the question because anything or anyone composed of information automatically becomes a creature of energy, space, and time. Such a creature could not avoid becoming a part of our universe (of space, time, matter, energy, entropy, and information) and would *not* be eternal.

Simply put, a God composed of (or describable by) information must become a part of *our* universe. Since our universe has a definite beginning (the big bang's singularity) and an eventual ending (the heat death of thermodynamics), such a God could not be eternal! Only a *zero*-information content God can escape being held captive by our material universe.

Here is a mystery worthy of great contemplation: how can God, who is not composed of information, remain the source of all the information that is becoming manifest in our universe? This very concept is so beautifully captured in the opening paragraph of St. John's gospel with the words, "In the beginning was the word, and the word was with God, and the word was God."

In the divine mathematics of God, we might say that God is both zero and infinity at the same time. All religions grapple with this problem in their own way. Some people (Jews and Muslims) say God is a single unity. We might say that in divine mathematics, for these people God is one. Other people (Christians and Hindus) say that God is simultaneously a unity and a diversity (a diversity of three for Christians and a diversity of infinity for Hindus). These religious people endorse the godly equation $1=3$ (for Christians) and $1=\infty$ (for Hindus). Like a Zen Buddhist Koan (i.e., an example: the sound of one hand clapping), the equations of godly mathematics challenge our rational minds and encourage us to set aside our everyday rational thoughts.

We now look more closely at some of the history of religious thinking. We will take a closer look at how our ideas about the truth of God can be better understood in light of the conclusions of the previous chapters. In most ways, traditional religious thinking is not really that far from the conclusions we are reaching in this work. However, some reinterpretations of traditional religious wisdom are necessary to find a unity of religious and scientific truths.

A Brief History of What Religious People Believe about God's Nature

During the fourth century CE, the new Christian church found itself face to face with a seemingly insurmountable problem. The Roman emperor Constantine, who had recently converted to Christianity, was planning to make Christianity the state religion of the Roman Empire. Since Constantine was an orderly man, he was eager to know exactly what Christians believed about God. This knowledge would enable Constantine to demand that his subjects adopt the right belief when they converted from paganism to Christianity.

However, when Constantine asked the numerous Christian bishops to state their own beliefs, he received many differing and confusing responses. In frustration, Constantine called a council of all the Christian bishops throughout the Roman Empire and demanded these bishops not leave the council hall until it was possible for them to speak with one voice concerning Christian beliefs. The first such council (held in the city of Nicaea[73] in what is now Turkey) gave rise to many other councils and synods, each becoming an event of the highest soul-searching controversy. The entire process of Christian belief definition took nearly one hundred years to complete. This period of Christian history proved to be a well-spring of Christian belief, initially spanning an entire spectrum of possibilities and finally focusing on a single set of orthodox beliefs.

This was a very painful period in the history of the new church, with many well-meaning bishops, priests, and deacons pointing fingers at each other and uttering horrible words like heretic, anathema, and excommunicated! However, in the end almost all Christians succeeded in closing ranks behind what theologians now call orthodox Trinitarian Christianity. The Trinitarians believed in a single monotheistic God with

---

[73] The Council of Nicaea was an early Christian church council held in 325 CE in the town of Nicaea (near what is now Istanbul) in Asia Minor. The theology of the Trinity was affirmed at this council, and its truth was recorded in a creed bearing the council's name.

three expressions called the Father, the Son, and the Holy Spirit. A perfect symbol for the Trinitarian God is an equilateral triangle possessing three equal legs and absolute rotational symmetry, repeating every 120 degrees.

We (the authors) find ourselves holding two divergent opinions about the Trinitarian God. First, of all the possible choices concerning God's nature that have come down to us from the fourth century, it is belief in the Trinitarian God that resonates most deeply within each of us. Second, we know that human beings cannot possibly hope to successfully describe or even begin to understand God's true nature because God exists beyond and apart from all space, time, energy, and information. Space, time, energy, and information are, of necessity (for we humans are creatures of information) essential ingredients in any possible description of God based on human logic and experience. Since as humans we are undeniably creatures of space, time, energy, and information, we simply cannot conceive of anything or anyone who exists apart from these ingredients.

Putting it simply, God's nature is fundamentally and completely beyond any and all human knowing. This is not to say that God's will is beyond our knowing; quite the opposite. The purpose of our lives is to discover God's will for each of us and then follow it. However, discovering God's will in our lives is not the same thing as knowing God's nature. Our true purpose is more along the lines of discovering our own nature and living out our nature to its fullest. We progress in our growth by attuning ourselves closer and closer to what God is calling us to do. We believe that the German theologian Karl Barth[74] was on the right track when he said, "God's being is in God's becoming."[75]

Perhaps the truth is that we do not so much know God as encounter God at moments of God's own choosing. God stands apart from our world of space, time, energy, and information, but through the mystery of quantum mechanics, God causes his will to manifest as information in our universe. What is there about God that we could possibly know? In truth, God is nothing and God is everything all at the same time. God is

---

[74]   Karl Barth (1886–1968) was a German theologian who was best known for his theology of revelation, which says that what we know of God is what God has chosen to reveal to us.

[75]   Eberhard Jungel and John Webster, *God's Being Is in Becoming* (New York: Continuum, 2004).

no thing, but God holds the potential for generating all things that could ever be.

To many, such a God feels simultaneously claustrophobic and explosive. However, the Bible gives us some insight at this point. In the book of Deuteronomy we read: "Then the Lord spoke to you out of the fire. You heard the sound of his words, but saw no form and there was only a voice" (4:12). God is described here as an unseen, formless voice speaking God's will to God's people. The God of Deuteronomy is described as formless (possessing a nature not describable by information) but capable of addressing God's will to God's people ("there was only a voice").

Our work suggests that God's voice becomes manifested in our universe in the moment when synchronous clusters of quantum events turn from being a coherent wave superposition into particles as their wave function collapses, thereby creating the presence of God's message (i.e., information). It is from this vast array of simultaneous wave function collapses that God's message becomes harmonized across space and time. Again in the words of the theologian Karl Barth, we can think of these coordinated manifestations of God's information as being "God's becoming." It is this field of space and time that contains the synchronized information from God to us that, for just a brief moment, becomes the sum totality of God's being.

We feel those images of God that are most compelling are those that are least sensible and least appealing to the rational human mind. Sensible images always convey an image containing some element of information for our senses and reason to act upon. But God is not composed of information! Yet God is the source of all information that flows into our universe. This concept represents the most profound of all mysteries.

How Fourth-Century Christian Theologians Learned What They Believed from Concepts of God They Could Not Believe

In the fourth century CE, there were several competing Christian theologies alongside the Trinitarians. The major fourth-century Christian heresies—as the Trinitarians called them—were the Arians[76] and the Gnostics.[77] The Arians (whose name has nothing to do with Hitler's "master race" but is derived from the name of a fourth-century Egyptian priest named Arius) believed in a hierarchy of gods, with God the Father being in charge, and God the Son and God the Holy Spirit being subordinate to God the Father.

The Gnostics also believed in a hierarchy of gods; however the Creator God, which the Gnostics associated with Yahweh, the God of the Old Testament, was thought to be evil. Therefore, for the Gnostics, all material things (including their own bodies) were likewise considered to be evil. The Gnostics believed that it was Jesus's role to save certain select humans from the clutches of this evil material universe. Ultimately, Jesus would deliver the souls of the selected ones to the highest God in the highest heaven who, unlike the Creator God, is good.

Jesus would first impart special knowledge (*gnosis*) to the believer to help facilitate the believer's transition from the material world to the heaven of highest good. The Gnostics were dualists who believed human beings possessed good souls but evil bodies. Practicing Gnostics either completely disregarded the needs of their bodies or indulged them wantonly. Gnostics believed in a hierarchy of heavens (including associated gods) and ranked them from the lowest level of evil to the highest level of the highest good.

We think that both the Arian God and the Gnostic God are far too worldly and reasonable in human terms (based on their hierarchical structures) for us to put much stock in them. Most of today's Christians believe in the Trinitarian God who (based on the Nicene Creed) is one God: "We believe in one God, the Father, the Almighty maker of heaven and earth." That unity is simultaneously a trinity, a diversity of three, that is equal and indivisible, called *hypostasis* in biblical Greek.

---

[76] The Arian heresy was an early Christian heresy that held God the Son and God the Spirit to be subordinate to God the Father.

[77] The Gnostic heresy was an early Christian heresy that taught that human bodies and all their functions are evil and only the human spirit is good. Gnostics believe that the earth and everything in it were created by an evil god and it was Jesus's mission to liberate those humans who possessed a special knowledge (gnosis) from the grip of this evil world.

Since it is impossible for any of us to get our logical minds wrapped around the Trinitarian God (unlike the god of the Arians and the god of the Gnostics), it is conceivable that the Trinitarians believe in a God who may indeed exist apart from space, time, energy, and information. In fact, God's unity within a diversity of three sounds a lot like the quantum mechanical concept of a superposition of unmanifested quantum wave functions, possessing the sense of *three-ness* within the totality of their wave function. For example, recall the double-slit Gedanken experiment discussed in chapter 2. More will be said about this correspondence a little later in this chapter.

The phenomenon of a quantum diversity of three occurs quite frequently in many atomic quantum wave functions because the mathematics that describe the wave functions operate in a three-dimensional space and naturally create a mathematical diversity of three. Such wave functions, like God's divine nature, must always exist as potentials with no information irreversibly recorded into our material world until wave function collapse. As we shall see, there is much food for serious thought within these analogies.

Christian theologians of the fifth century CE went through a very similar soul-searching process while wrestling with the question "Exactly who was this man we call Jesus?" A whole spectrum of beliefs concerning the nature of Jesus had emerged during the first four hundred years of Christian church history. Some churchmen thought Jesus was truly God and had only pretended to be human. This belief was called *docetism*, and its root word means "playing a role"). Other churchmen thought Jesus had been completely human and was only later adopted by God because of the perfect life he led. This belief was called *Adoptionism*. Both of these beliefs were eventually discredited and discarded by the Church in favor of a general agreement that Jesus was simultaneously both fully human and fully God.

St Athanasius[78] famously said, "God had become man so that man might become God."[79] This is a powerful and profound statement. It has served as food for thought for many centuries. The first half of the

---

[78] St. Athanasius (296–373 CE) was an early Christian theologian who was a strong supporter of Trinitarian theology and an opponent of the Arian heresy.

[79] *Catechism of the Roman Catholic Church*, Part 1, Section 2, Chapter 2, Article 3, Para. 1.

statement surely refers to Jesus's incarnation into human form, but how does man become God? Perhaps God in the form of the man Jesus offers us examples of this possibility. Perhaps it was Jesus's mission on earth to awaken within each of us the God nature residing within our hearts (i.e., the divine information God is constantly incarnating within us).

The concept of Jesus's simultaneous divine and human natures should sound very familiar based on our exploration of quantum mechanics. Jesus's simultaneous human and divine natures correspond directly to what the founders of quantum mechanics concluded about the simultaneous wave and particle natures of all matter and energy (which can never be separated or broken down). Of course, the conclusion that Jesus is simultaneously human and divine violates all rational human logic and common sense—perhaps to the advantage of the believer.

The God of the World's Religions

Next we will take the time to briefly explore what the adherents of some non-Christian religions believe about God's nature. The reason for this very brief exploration into non-Christian world religions is to show how these religious traditions harmonize within a unified truth. The harmony is in the finding of the truth. We hope that followers of these religions will also feel called to pursue the same kind of investigation within their faith that we have here conducted for Christianity.

Hinduism

First we turn our attention eastward to Hinduism, the religion of India. Hindus say they have 330 million gods. We think that by making this statement, Hindus may be saying that God presents an infinite variety of faces to humanity. We humans have the ability to perceive these faces and distinguish one from the other. In Hinduism there is a principle which translates roughly from Sanskrit as "divine essence." The meaning of divine essence is roughly this: it is essential to believe that God resides within each one of these essences and is equally available to each of us regardless which one of the many Hindu gods and goddesses we might choose to devote ourselves to. Each practicing Hindu typically finds one special deity that in some way matches his or her own personal predisposition and personality. By devoting him or herself to this deity, a believing Hindu perceives the essence of the totality of God working within his or her life. It doesn't matter which deity is chosen. The point is to devote oneself to find that particular essence of God that shines most brightly and with the highest meaning into your life.

Although there may be a countless number of Hindu deities, all of them are part of a great hierarchical tree of deities that originate from Hinduism's three principal faces of God (there is that *three* again). All lesser Hindu deities are in some way related to the three principal deities, either as their aspects or as their avatars (incarnations). The existence of these three principal Hindu deities is very interesting given the profound similarity between this fundamental three-*ness* of Hinduism and the three "persons" of the Christian Trinitarian God.

The Hindu "Trinity" consists of Brahma, Vishnu, and Shiva. Brahma is the god of creation, Vishnu is the god of preservation and the lawgiver, and Shiva is the god of distruction, the distroyer of all things and all beings. Hindus believe that Brahman (not to be confused with Brahma), the supreme cosmic spirit, resides within each one of us as "soul," called the Atman. It is understood by Hindus that Atman exactly equals Brahman. The deity within is *exactly* the same as the deity without. Therefore, most Hindus seek God first within themselves rather than seeking God in the outside world. Mahatma Gandhi discerned the same phenomenon in the communities of Jews, Christians, and Muslims. Moses, Christ, and Mohammad manifested to Gandhi the same holy name. The beliefs of Hinduism may seem complex and convoluted to the Western mind. However, Hindus are of the opinion that their religion encompasses all of spirituality. Hindus say that if something is truly of God, then it has a place within the ancient, accepting, and flexible religious framework of Hinduism.

Buddhism

Buddhism was born from Hinduism in much the same way Christianity came from Judaism. Buddhism goes to the other extreme from Hinduism in its understanding of God's nature. Buddhism has no place for God or for gods. Buddhism is a religion of cosmic process, of cycles of life, and of the evolution of humanity toward its ultimate destiny. The goal of Buddhism is nothing short of achieving a final end to the pain and suffering of each human life.

The Buddha, a man who lived in India about four hundred years before Jesus lived in Palestine, meditated for many years on the meaning and causes of pain and suffering. The answer came to him at the moment of his enlightenment. At this moment, the Buddha realized our pain is caused by our attachments—nothing more and nothing less. Enlightenment, which the Buddha said anyone can achieve, grows out of a recognition that we torture ourselves by attaching ourselves to all the things, people, goals, ideas, and "should and ought" that act in and upon our lives.

The list of attachments is lengthy. Those who aspire to enlightenment can clearly see that their attachments weigh them down, causing their spirits to be impeded in their progress along the path of spiritual evolution. Where was this evolutionary path leading? The Buddha's answer was to teach his disciples that all paths lead to a state called Nirvana. Nirvana is the condition of no action, no attachments, no form, no movement, and no change. A human spirit reaches Nirvana through a process of enlightenment, which is achieved by embracing a total annihilation of all forms and complexity that once constituted the individual's life. By achieving Nirvana, an individual becomes totally merged into the universal energy of the cosmos, in which individuality is completely lost. The soul no longer desires to regain it.

The Buddhist nonpicture of God's nature is so un-sensible and illogical that we (the authors) feel Buddhists are definitely onto something. If we equate form with information and attachments with choice, it becomes very easy to draw parallels between the Buddhist concept of nonattachment (leading to the formlessness of Nirvana) and the concept of a God residing beyond and apart from our world of information. This "God" (assuming we equate the Buddhist Nirvana with the Christian concept of God) is somewhere beyond all of the unending choices that characterize our world of information.

We humans are compelled to choose; that is the rule of our human existence. We are tempted to say that a Buddhist Nirvana and a God who exists beyond space, time, and information are one and the same. To successfully draw this parallel, we must first accept that a state of Nirvana and a being God can be one and the same. This striking parallel takes concepts from science, such as quantum mechanics and information theory, and develops them so their wisdom comes back full circle to be in alignment with one of the oldest and best known of the world's religious philosophies.

Taoism

Now let us consider one more world religious tradition: Taoism. Let us briefly examine how its concepts of truth resonate with the truth discerned by way of the scientific method. Taoism originated and developed as a philosophy of life among the sages of ancient China. The key teaching of these sages is that within our natural world, all forces for change act through the continuous transformation of polar opposites, one into the other. These changes are played out in many different ways (e.g., light changes into darkness; male and female merge and change roles; the strong and the weak exchange their roles; hot turns cold; cold turns hot).

Change is the only constant in this world. The ancient sages of China recognized that only one thing remained forever changeless, and they called this one essence of changelessness the Tao. Even the name Tao (like the name God) is simply a placeholder because the real Tao can never be named. The Tao was never composed of pairs of changing polar opposites (like God, who is not composed of information), unlike everything else in our universe that is, so the Tao must exist apart from this universe of form and change.

The Tao is eternal while everything else in our universe is temporary (including the universe itself!). Change will eventually transform all things that are composed of pairs of polar opposites, but not the Tao. The Tao remains constant and changeless for all time, from before the beginning of time (the big bang?) till after eternity has brought an end to our universe (heat death?). The Tao never changes; the Tao always remains the same. There is "no-thing" within the Tao, because all things change, only the no-thing of the Tao remains eternally unchanging, forever the same.

Taoism and Nature

Of all the world's religions, Taoism resonates most directly with the picture of God that emerges from our analysis of the unity of religious truth with the truth of the natural sciences. This is perhaps not surprising, because the ancient Chinese sages who formulated Taoism regarded themselves as serious students of nature. If these sages were alive today they would, one hopes, view this work as a worthy extension of their own explorations into the spirit.

The great intuitive insight of these Taoist sages into the true nature of the world around us cannot be overstated. Taoist art is filled with scenes from nature: flowing brooks and swimming fish, birds in the trees alongside the river banks—and all against the background of impenetrable sameness appearing as fog or as cloudless sky.

Change occurs everywhere in nature as all things obey the ebb and flow of their roles in life's eternal dance. The Tao always remains unchanging at the center of this celestial dance. Tao is calmness itself, and it resides at the very center of a frantically changing pattern that is all life. Tao is like the eye of a hurricane, where storm winds are blowing all around but all is calm and still at the very center.

It is key to our understanding that God must be beyond all bounds of space, time, and information. This understanding is in complete harmony with Taoism. The similarities are especially profound if one accepts as a one-to-one correspondence the choices required in determining each bit of information on one hand and the Taoist concept on the other (which holds that nature is constructed out of pairs of dynamically changing polar opposites). This central understanding of the unity of science and religion associates the world of the changing polar opposites of Taoism with our natural thermodynamic world of energy and information. The understanding we have reached in this work places God beyond all information content in a way that is completely consistent with the Taoist insight that the Tao exists beyond the world of dynamic changes.

Returning to Thermodynamics

Finally we direct our thoughts and meditations on God's nature to the question of the ultimate purpose. We have learned from the sciences of thermodynamics and cosmology that our universe is proceeding outward from its beginnings in the big bang toward its final state, an infinitely fine distribution of vanishing, low-density matter and energy existing in a state of total chaos and disorder called the heat death.

The heat death will constitute our universe's final chapter. The process moves from infinitesimal space (the big bang singularity at the moment of creation) to nearly infinite space (associated with the heat death). The progress is accompanied by the rise and fall of vast quantities of information that are first generated and then lost to entropy (i.e., the lost or hidden information of chaos) all along the way. All of the possible information scenarios are given a chance to play out over the vast timeline that extends from the big bang singularity to the heat death.

As this evolutionary process continues, the universe moves along the arrow of time toward its ultimate destiny and fate. Like everyone and everything in the universe, we also play our role in this process, recording the information of our stories. They live on for a while but in the end are extinguished. Each one's story is inevitably returned to the entropy of un-knowledge at thermal equilibrium. As Shakespeare had Prospero explain it, "The cloud-capped towers, gorgeous palaces, the solemn temples, the great globe itself, yea all that it inherit, shall dissolve . . ."[80]

This return to thermal equilibrium is experienced by all information bearing systems, including ourselves. What scientists refer to as the heat death is simply the final stage in our universe's existence as it ultimately plays out all possible information-generating scenarios. In this final state, all form has been extinguished, information has returned to entropy, and any further change has been rendered impossible. Information has ceased to exist, time has stopped, and the space enveloping the universe has reached its ultimate expansion. Everything has stopped, and all is perfectly still. Every direction in space is indistinguishable from every other direction, and time, the final measure of change, has ceased. The entire universe has become a vast, sandy desert. Life's final story has been told. The last book has been written. All that has ever been has now changed into finely granulated dust.

But the universal laws of the conservation of matter and energy still hold sway! The amount of matter and energy in the universe's final state will be exactly equal to the total amount of matter and energy that existed in the big bang singularity at the universe's birth. This fact of life eternal places upon us the duty to find our present life's assignment, striving to live that to its fulfillment. That is the ethical component of our days on earth. That is to say, it is the duty of all of us who are alive only briefly to live out our lives to the fullest in accordance with God's plan for our lives. It is up to each of us, to the best of our ability, to embody and carry out the plan of God for our lives as we each express our unique talents and gifts.

---

[80] William Shakespeare, *The Tempest,* Act IV, Scene 1.

From Nicaea to Copenhagen: A Juxtaposition of Quantum Mechanics and Orthodox Christian Beliefs

As you are about to see, the formal framework of quantum mechanics and the formal framework of orthodox Trinitarian Christian belief are remarkably similar. This situation makes one wonder if the founders of quantum mechanics had a Christian theological predisposition when they developed the science. That, however, seems rather unlikely since most of the founders of quantum mechanics were Jewish, and secular at that. (Einstein's father would talk to a young Albert about an ancient superstition called religion.)

It is equally impossible that fourth-century Christian theologians had quantum mechanics in mind as they wrestled with questions concerning God's nature, since it would be almost two millennia before quantum mechanics would be developed. It is highly unlikely that there is anything contrived about these similarities. The only other person who has recognized and stated these similarities is John Pokinghorn,[81] who discussed the relationship between quantum mechanics and Christianity during his 2003 Terry Lectures.

It would seem, given all of the evidence, that somehow these fourth-century theologians developed by pure intuition and inspiration a description of the nature of God that bears a remarkable resemblance to a description of the natural world developed by a branch of science that would not be founded for another two millennia. There has to be more to this picture than meets the eye.

---

[81] John Pokinghorn (born 1930) is an English physicist and Anglican priest. Pokinghorn has written extensively about the convergence of science and religion.

Nicaea, 350 CE

As we have already seen, when the emperor Constantine[82] converted to Christianity, he, being an emperor, wanted to know exactly what the people of his empire were to believe when he declared Christianity to be the state religion of Rome. To this end, he demanded that all Christian bishops within the empire meet in a council at the city of Nicaea, which is located in what is now Turkey (near the city of Istanbul). The purpose of this council was to give the bishops an opportunity to reach an agreement on what all Christians must believe.

At the time, Christian belief was seriously fragmented. The major issues dividing Christians had little to do with Jesus's teaching and even less to do with the Hebrew traditions of the Old Testament. The real issues that divided Christians focused squarely and completely on the very question of who really is God. For Christians, God reveals himself to humans in three ways. The first way is through the revelation in the life and teachings of Jesus. The second way is through the Father God of the Hebrew Old Testament. The third and final way happens when God speaks directly to an individual's heart in the form of what Christians call the Holy Spirit.

All Christians in the fourth century agreed some kind of fundamental three-ness about God had to exist, but the question of how human beings must think about these three seemingly separate expressions of the one God separated Christians. Do these faces of God represent three separate Gods? Do they form an hierarchical corporate God, with one God expression being of higher authority than the other two? One school of thought, the heresy of modal Sabellianism,[83] assumed one of these faces was presented part of the time and at will, God assumed a different face whenever God wished.

Also, the very nature of the man Jesus was a source of constant debate. Some Christians felt Jesus was a human being, just like any other human being, who happened to have a very special relationship with God

---

[82]   Constantine (272–337 CE) was a Roman emperor who changed the state religion of Rome to Christianity in 313 CE. It was Constantine who called the council of Nicaea.

[83]   The modal Sabellianism heresy was an early Christian heresy that held God could choose to change into the mode of Father, or Son, or Holy Spirit at will.

(Adoptionism). Others felt Jesus was truly God going around planet earth in a "man suit" (Docetism).

These disagreements were endless and often pitted Christian against Christian in sometimes very unchristian ways. Sects were formed based on the various beliefs, and their adherents would have nothing to do with each other. In some cases, these bad feelings went so deep that Christians subscribing to one set of beliefs were no longer welcome at the churches of those committed to an opposing position. It was this atmosphere of Christian diversity and disagreement that immediately preceded (and in some cases followed) the council at Nicaea. It is truly a miracle that any common agreement on beliefs was ever reached by this group of bishops, considering how fragmented Christian beliefs had become. (Perhaps the emperor motivated the bishops by withholding their food until the work was completed!)

In any event, a document called the Nicene Creed (this creed was later refined at the council of Constantinople a half century later) did emerge from the council of Nicaea. The Nicene Creed has come to be thought of by the majority of the world's Christians as the sole acceptable statement of orthodox Christian belief. According to the final draft of the Nicene Creed, God is truly a unity of God-ness. God is one God, complete unto God's self. God is eternal and omnipotent. God is truly the one and only God. There is no other God but God. However, God has a three-ness to God's nature.

God was expressed to humans first, but perhaps not foremost, as God the Father of the Old Testament (whom Jesus prayed to). God is also revealed to humans in the life and teaching of the man Jesus, who is also called God the Son. God is also experienced by human beings in moments of inspiration and grace as God the Holy Spirit. Therefore, God is a three-ness. God is not three gods; God is always one God but with three expressions. The God who is simultaneously one God with three expressions is called the Trinity.

As mind-boggling as it may be for us humans to contemplate, the God of the Nicene Creed is a oneness expressed within a three-ness. This means God is always one but God is also always three, which is to say that 1=3 in the divine mathematics that is completely baffling to human logic! A later clarifying creed (called the Saint Athanasius Creed), using a wording that wields all of the precision of a geometric proof, painstakingly insists that God the Father is eternal and omnipotent,

God the Son is also eternal and omnipotent, and God the Holy Spirit is also eternal and omnipotent, *but* these three Divine expressions are *not* in any way three separate gods. God is first, foremost, and always one God who is eternal and omnipotent.

Many years later, yet another council was held in the city of Chalcedon, which is close to Nicaea. At this council the nature of Jesus was seriously debated. This council concluded that Jesus was simultaneously a fully human being (just as human as you and me) and at the same time fully God, which is to say the incarnated (i.e., having taken human form) God the Son.

A few short years later, Saint Augustine,[84] bishop of Hippo (in north Africa), after further reflections on the Trinity, reached the conclusion that the relationships among the three-nesses of God are what make God truly accessible to human beings. For instance, the Father loves the Son, the Son is beloved by the Father, and it is the Holy Spirit who is that love that passes between the Father and the Son. It is God the Holy Spirit who teaches the human heart to love. Augustine felt it is these relationships within the three-nesses of God that create a profound awareness of God within each of us. Augustine felt that humans learn first and foremost about God and God's goodness through God's relationship to God's self!

There is a beautiful legend that one day Bishop Augustine was walking by a beach near his home while he was contemplating on the meaning of the Trinity. As he walked, he saw a small boy playing at the water's edge. As the bishop drew closer, he saw that the boy was digging a hole in the sand in such a way that as the waves hit the beach, they would drive water into the hole. Bishop Augustine asked the boy what he was doing, and the boy replied he was digging a hole that would hold all of the water in the ocean. The bishop replied that it would take a very long time for all of the ocean's water to pour into that hole. The boy replied, "I will have the entire ocean in my little hole long before you even begin to understand the Trinity!" Generations of Trinitarian scholars and theologians have heeded these words.[85]

---

[84] St. Augustine of Hippo (354–430 CE) was an early Christian theologian. Augustine felt the real essence of the Trinity was to be found in the relationship of its persons.

[85] The legend of St. Augustine and the boy on the beach is beautifully depicted in the grand blue window over the entrance of St. Augustine Catholic Church in

Confusion over Language

Unfortunately over the years, centuries, and millennia, much confusion has arisen over how best to express the God of the Trinity in the words of language. By way of mistranslation from the original ancient Greek into Latin, the three-ness of God has come down to our time as poorly translated into English as persons. This happened because the original Greek word for God's three-ness is *hypostasis*, which was confusingly translated into Latin as *persona* (meaning masks). Later this confusion was intensified by translating persona into English as persons. Persons makes no sense at all as a word for representing God's diversity of three and is really terribly confusing because the word persons conveys the false message of God's humanness. Persons completely loses the original intent of the Greek word hypostasis, which carried a meaning of God's three-ness in the midst of God's eternal and omnipotent oneness.

A similar language problem occurred with the word for God's oneness, *homoousios*. In ancient Greek, the word homoousios was used to convey a sense of unity and oneness. Unfortunately, homoousios was translated into Latin as *substantia*, which was later translated into English as substance. This poor translation of the ancient Greek word homoousios into the English word substance has resulted in a complete misunderstanding of the meaning of God's oneness. The word substance implies that God's oneness is in fact composed of something material! Both science and religion should be crying foul at this gross misuse of language and bemoaning the numerous misunderstandings this misuse of language has generated.

The Christian doctrine of the Trinity remained relatively unmodified for well over a millennium. However, in the mid-twentieth century, a German theologian named Karl Rahner[86] developed a concept of an immanent Trinity that is separate from an economic Trinity. The immanent Trinity, as envisioned by Rahner, contained the totality of God, complete

---

Waikiki, Oahu, Hawaii. A brochure in the narthex of the church tells the story of the legend.

[86] Karl Rahner (1904–1984) was a German Jesuit theologian who developed the concept of an immanent Trinity (that of God that God chooses not share with humans) and an economic Trinity (that of God that God chooses to share with humans).

in every way. However, Rahner goes on to say that the economic Trinity contains only that portion (whatever a portion of God can possibly mean) of God that God chooses to *reveal* to human beings.

By thinking along these lines, it is clear that human beings are not privileged to experience the totality of God that is contained within the immanent Trinity. What humans experience is that portion of God (the economic Trinity) that is responsible for the presence of God in our lives as we read the Holy Scripture and allow the Holy Spirit to speak to each of us in unique ways. In a more general sense, we could say that it is the economic Trinity that interacts with our material universe in a role of our universe's architect and sustainer.

In all ways it is the economic Trinity that causes new information to appear within our material universe, thus defining God's plan for our universe and everything and everyone within it. God's divine information could take the form of blueprints for constructing entire galaxies or a chemical formula for a giant organic molecule that might lead to the synthesis of self-replicating life-forms. It could also take the form of knowledge of good and evil within each of us. However, as we have previously discussed, information in all its various forms is a purely thermodynamic property of the material universe.

God, who exists apart from this universe of matter, energy, information, and thermodynamics, cannot be composed of or contained within the material. However, if human beings are to be aware of God's existence and God's plan for our lives, it is absolutely essential for God, in some way, to introduce the information of God's plan into our material universe. Without the process of manifesting the message that God is there to make us aware of, there could be no way for human beings to know of God's existence. Therefore, based on the laws of physics and information, it is absolutely essential to preserve the unity of truth that it be the economic Trinity that causes the information God wishes to share with us to come into material existence in a way that does not violate any of the laws of science (i.e., conservation of energy, etc.). There must be a means by which God writes out God's plan for us in a form of information we humans can recognize and make use of.

On to Copenhagen, 1927 CE

In 1905, Albert Einstein was an obscure patent agent in Bern, Switzerland. In his spare time, he wrote a number of scientific papers that turned the world of physics upside down! Two of these papers erased any doubt concerning the reality of atoms, and a third laid the foundation for Einstein's ongoing theories of relativity. A fourth paper, which would later win Einstein a Nobel Prize, theoretically explained the findings of an obscure physics experiment called the photoelectric effect. In this paper, Einstein conjectured that light must indeed be composed of particles, although less than thirty years earlier, the great Scottish scientist James Clerk Maxwell (of Maxwell's demon fame) had demonstrated to everyone's satisfaction that light could only be understood as a wave phenomenon.

Could both Maxwell and Einstein be right? As it turned out, yes, they both were right. The only realistic conclusion science could draw from Einstein's revolutionary concept was to accept as fact that light can simultaneously exist as both a particle and a wave! Within two years, a young French graduate student named Louis de Broglie hypothesized that if light waves can behave as particles, then perhaps particles of matter (such as electrons and protons) could also behave as waves. Very soon experimental confirmation of de Broglie's theory became available to the world of science, and a new branch of physics called quantum mechanics was born out of the mystery of how matter and energy could exist simultaneously as both waves and particles.

As we have already seen, the Danish physicist Niels Bohr took up the quest to understand the meaning of quantum mechanics from Einstein and de Broglie. In the process, Bohr produced a theoretical explanation for the internal workings of the atom. Bohr's atomic theory involved the concept that electrons circle about the atom's positively charged nucleus and do so within highly prescribed and discrete sets of orbits. This is a direct result of the wave-nature of electrons. These levels could be calculable based on the still-emerging mathematical understanding of quantum mechanics. According to Bohr, atomic electrons could only exist in certain well-defined allowable orbital states. All other electronic orbits were expressly forbidden by nature!

Bohr's atomic theory was wildly successful, both in term of its ability to explain experimental results and more philosophically in its ability to bring about greater understanding of the nature of subatomic reality based on the mathematics of quantum mechanics. Riding high on this wave of his success, Bohr founded an institute for theoretical physics in his home city of Copenhagen, Denmark. He surrounded himself with many of the brilliant young scientists of his time, some of whom would later serve as architects of the emerging science of quantum mechanics. This school of quantum scientists, and the theories produced by its members, came to be known collectively as the Copenhagen school (and interpretation) of quantum mechanics.

Niels Bohr was not only a brilliant scientist; he was also a very good philosopher. It was his love of philosophy and his personal search for truth that drove his and his institute's scientific quests. However, not everyone agreed with Bohr's theories and their philosophic interpretations. Albert Einstein, although he was one of the founding fathers of quantum mechanics, became very skeptical of the direction quantum mechanics was heading under Bohr's leadership.

Over the ten-year period from 1925 to 1935, a rivalry developed between these two great scientists. However, because Einstein was unwilling or unable to produce viable alternatives to Bohr's theories, his role throughout this rivalry remained largely that of a gadfly. Although Einstein continued to "bug" Bohr deeply, he did play an important role in keeping Bohr on the right path by forcing him to deal with a number of very difficult conceptual issues.

Bohr's understanding of quantum mechanics drifted (after insistent prodding from Einstein) more and more in a direction of viewing the wave function (the mathematical entity describing any kind of quantum phenomenon) as being less than a creature of our real world and in all likelihood a kind of mathematical abstraction whose usefulness expires (at wave-function collapse) when a real laboratory measurement takes place. For instance, the Copenhagen school would deny that an electron—any electron, for that matter—was real in a physical sense until the electron was experimentally measured by some kind of laboratory apparatus that was capable of permanently recording the presence of the electron!

As this line of thinking continued to be fruitful, Bohr found himself coming more and more to the philosophical conclusion that it is always the measurement process itself that confers reality on any quantum mechanical entity. Einstein, who was having a great deal of difficulty with Bohr's seeming denial of the reality of unmeasured quantum entities, once remarked in total frustration, "Professor Bohr, are you telling me the moon doesn't exist until I look at it?" Bohr's response was never recorded.[87]

However, in spite of Einstein's best attempts to discredit Bohr's quantum theories, the basic structure of Bohr's approach to understanding the subatomic quantum world remained strong and unassailable. It turned out that in a strange sort of way, Einstein's criticisms were counterproductive to his cause, in so far as they drove Bohr to come up with new and successful retorts that were, to Einstein, even more bizarre and unacceptable. In the final analysis, the Copenhagen interpretation of quantum mechanics, by way of the Bohr-Einstein dialog of criticism and rethinking, came to include the following key points about the reality of all subatomic phenomena:

1. There exists a wave-particle duality in all subatomic entities (matter and energy). True understanding of the subatomic world demands that an entity's wave nature and particle nature be simultaneously considered.

2. Any and all attempts to prove that an entity is solely a particle will fail on account of the entity's wave nature, and any and all attempts to prove that an entity is solely a wave will fail on account of the entity's particle nature.

3. The wave function of a quantum mechanical entity can be calculated mathematically but can never be measured in the laboratory. The only way to physically understand the wave function is to regard it as a statistical quantity predicting a probability that its quantum entity will assume, upon measurement, a specific measurable attribute, such as a position or a momentum.

4. Not all attributes (dynamic variables) of a quantum entity can be measured simultaneously with complete precision. According to the (Heisenberg) uncertainty principle, the more precisely we measure an electron's position, the less precisely we are *able* to measure the electron's momentum. In the same way, the more

---

[87]   Manjit Kumar, *Quantum, Einstein, Bohr, and the Great Debate about the Nature of Reality* (New York: W. W. Norton and Company, 2008), 352.

precisely we measure an electron's energy, the less precisely we are *able* to measure an electron's time of arrival.

5. Because of the universality of the uncertainty principle, it is impossible to know with complete precision a set of initial conditions for any quantum entity. This lack of initial conditions means causality, based on a deterministic evolution from a set of initial conditions, is simply not workable within the quantum mechanical framework of understanding.

6. An entity's wave function is composed of a superposition of all possible wave function options that might describe the entity's future evolutionary path.

7. The detailed behavior of a quantum entity just prior to measurement is describable by the wave interference effects (i.e., coherence) among its various superpositions.

8. The wave function instantly ceases to exist (wave function collapse) when a measurement takes place. All of the wave function's superpositions and their coherent interactions instantly cease to exist at the moment of measurement. At the moment of measurement, the entity assumes its particle nature (with all of its attributes concentrated at a single point in space-time).

9. Everything that can ever possibly be known about an entity's attributes is contained within its wave function. However, the number of attributes that can be simultaneously measured is entirely dependent upon the details of the experimental measurement system and on the restrictions of the uncertainty principle. Under a given set of measurement conditions, those attributes that can be measured constitute the totality of what (in that moment) constitutes reality for that entity. Only those attributes that can be measured make up the sum total of the entity's reality. Simultaneous knowledge of all of an entity's dynamic variables is simply not possible within our real world.

10. Action at a distance is possible for two (or more) quantum entities that were created together and remain entangled for all time. If the attributes of the combined entangled entities are known, then by measuring either of the entities at a later time, the other entity is simultaneously measured instantly, no matter how far apart the two entities have become.

Corresponding Orthodox Christian Beliefs with the Principles of Quantum Mechanics

There are many striking similarities between the beliefs of Trinitarian Christianity and the Copenhagen interpretation of quantum mechanics. We can only guess why this is. Perhaps the fourth-century Christian theologians who codified these beliefs intuited a Trinitarian God whose structure closely matches the process by which God's divine information is revealed to believers as they pray and study the Bible. Perhaps the answer is simply that both quantum mechanics and Trinitarian theology describe the way in which God chooses to communicate with our material universe and with ourselves.

Below is a list of correspondences between the Copenhagen interpretation of quantum mechanics and statements of orthodox Trinitarian Christian belief (based on the conclusions of the fourth-century councils of Nicaea and Chalcedon). We assume these statements of Trinitarian belief apply to the economic Trinity and perhaps to the Trinity as a whole.

1. God is eternal. This belief in God being eternal corresponds to a quantum wave function prior to a measurement. Since a quantum entity prior to measurement is in a state of what can only be described as unreality, the quantum entity itself will go on forever in its present state until a measurement disturbs it. For example, the wave-function of atomic electrons that surround the nucleus of any atoms remains unchanged for all eternity if no measurement is ever made. A second example is the wave function of a DNA molecule that can potentially store genetic information all the way back to the first life on earth.

2. God is omnipotent. This belief corresponds to the statistical nature of the quantum wave function, which gives human observers only a probabilistic knowledge of the behaviors of quantum entities but no absolute knowledge of the behavior of a single entity being measured at some particular point in time. Assuming that it is God who controls the final outcome of each quantum mechanical measurement, God's abilities in this regard are truly omnipotent. Here also the principle of entanglement implies that whoever measures one entity of an entangled pair of entities has simultaneously measured the other. Since we are assuming that it is God who determines the final outcome of these measurements, clearly based on quantum entanglement, time and space are no barriers to God's determinations of final measurement outcomes.

3. God is one God (homoousios). The oneness of God corresponds to the final and irreversible quantum measurement that occurs each time a quantum entity is measured by an experiment that brings into existence new and unique information about this single entity's attributes. The best predictions human beings can make are statistical predictions of an entity's attributes averaged over an ensemble of identical measurements. However, it is only through God's oneness that uniquely defined information springs into existence at the moment of a quantum mechanical measurement.

4. God's diversity (three hypostasis). God's diversity corresponds to the diverse superpositions of a quantum wave function. However, the diversity of the Trinity is three, and the diversity of quantum superpositions is in principle limitless. Perhaps it is our own human consciousness that is uniquely tuned to God's diversity of three. On the other hand, a diversity of three may be unique to the Christian faith, while other religions (i.e., Hinduism) are very comfortable with a God of unlimited diversity. Another factor worth considering is that the diversity of three corresponds to the three spatial dimensions of our universe. Many quantum wave-functions that involve spherical structures, such as atoms, have wave-functions with a diversity of three.

5. God's diversity within God's unity corresponds to the many (i.e., diverse) quantum mechanical wave function superpositions adding together coherently to determine the final probability of a quantum measurement.

6. It is the relationships among God's diversities that teach human beings about God's virtues (this was first taught by Augustine of Hippo in the fifth century CE). The relationships among God's diversities correspond to the coherent wave interference that occurs among the various wave function superpositions that interfere with each other prior to quantum measurement. (See the discussion of the double-slit experiment in chapter 2.) This unique and irreversible situation is the result of measurement and corresponds to Augustine's divine virtues (love, truth, etc.) being received by humanity.

7. The economic Trinity is that of God that God chooses to share with humanity. The immanent Trinity is that of God that is God's business alone. The economic Trinity corresponds to quantum mechanical measurement because it is by way of the economic Trinity that God shares God's plan with us. God creates information (as a result of the quantum measurement process) as a way of manifesting God's plan within our material world.

8. God's incarnation as the man Jesus, in which Jesus is simultaneously fully human and fully God, corresponds to the quantum mechanical wave-particle duality. There is no separating a quantum entity's particle nature from its wave nature. Any experiment intended to prove that an entity has a purely wave nature will fail because of its particle nature. Conversely, any experiment intended to prove that an entity has a purely particle nature will fail because of its wave nature. So it is with Jesus. Jesus's God nature and Jesus's human nature cannot be separated or subdivided in any way.

9. Since God is beyond space, time, and information, God is not something or someone we humans can possibly understand given

our limited, information-based human intelligence. What we really do is experience God in those moments when God chooses to meet us at the crossroads. God's revelations at the crossroads correspond to the synchronistic summation of all those quantum measurements that are being simultaneously perceived by us in a moment of revelation.

10. Prayer, as a process, corresponds to a quantum measurement. Like a quantum measurement, prayer has the ability to confer reality on God's presence in our lives. The question or partition that is being presented by the one who is praying corresponds to a laboratory equipment setup for a quantum measurement (e.g., Can the single-slit experiment resolve the question? Can an electron ever act purely as a particle?). In correspondence to the statistical nature of quantum measurements, the answers to our prayers may come in unexpected forms and/or at unexpected times.

Based on this list of correspondences, we conclude that the framework of quantum mechanical behaviors operate in much the same way as the framework of the economic Trinity. We propose the meaning of these correspondences is derived from an understanding that quantum mechanics provides a mechanism by which the economic Trinity communicates God's plan into our material world. To keep our thinking consistent, we furthermore propose that the economic Trinity is that of God that speaks directly to our own human consciousness. In this way, we might refer to our personal experience of God as the macroeconomic Trinity. Alternatively we could say that the process by which the economic Trinity causes information to become manifested in our universe at the subatomic level (by virtue of quantum mechanical interactions) be called the microeconomic Trinity.

It is by virtue of God's extremely subtle communication abilities involving the coordination and timing of very large numbers of quantum events that our universe and every person and thing within our universe take their form and are granted their existence. It is at this cutting edge of reality that the truth of God, as understood by the world's religions, and the truth of the natural universe as understood by science (through the wisdom of quantum mechanics, thermodynamics, and information theory) come together in perfect alignment and agreement. The truth, by virtue of this alignment, is unified and complete. In the next chapter, we will investigate how this unified truth reaches deep into the lives of all living things.

CHAPTER 6

LIFE

There has been life on planet earth for at least the last five billion years. How life on earth began is one of the greatest of all mysteries. Biologists like to speak of a small, warm pond where the molecular ingredients of life somehow combined in ways that produced the self-replicating molecular structures of early life. However, all scientific attempts at duplicating these conditions in the laboratory have failed to produce anything more lifelike than gray-black muck. There is no question that whatever is the essence of life does not just fall out of the chemistry of these elements. A debate that has raged for centuries (and continues to rage in our day) is whether life emerged spontaneously and naturally on its own accord or if the emergence of life required the intervention of a creator God. Let us consider these questions from our newfound point of view that God is constantly in communications with our universe via the information channels that are made available by the unpredictability of quantum mechanical entities.

Life Blooms

Good morning, star shine. Did you know that we are all stardust? Shortly after the big bang, after clouds of photons and hailstorms of subatomic particles had blazed forth, extremely large stars were formed shortly after a period of the universe's development cosmologists call inflation.

Our sun is a very small star when compared with the stars of the early universe. Because of its smallness, our sun, like most of the other modern stars, is only capable of manufacturing one new element by fusing hydrogen atoms up to the element helium (next element on the periodic chart). But the ancient, giant stars of the early universe were much larger than our sun, and because of their great size and extremely high internal temperatures, these early stars were capable of acting as breeding grounds for all of the elements on the periodic chart. In particular, the element carbon, which is so essential for life on earth, was produced in the stellar fusion furnaces of these ancient stars, making it possible for us billions of years later to be writing this book and you, the reader, to be reading it. It is a fact that without carbon there would be no you and no me since carbon is an absolutely essential ingredient for life.

Our earth was formed about 6 billion years ago out of the cosmic stardust left over from the giant stars of the early universe. This dust was the product of supernova explosions that in time consumed most of these ancient giants. Fortunately for us and for all life on earth, our earth received a good supply of every element in the periodic chart, including the essential element carbon, the very stuff of life.

Carbon, hydrogen, oxygen, nitrogen, and phosphorus are the key ingredients of life on earth. Carbon atoms are capable of combining with hydrogen, oxygen, nitrogen, and phosphorus atoms to form the large organic molecules that make up all living things. Some molecular biologists think rock crystals may have acted as a template or patterning matrix for early life, possibly serving as the catalyst for the formation of these early organic molecules. Carbon is such a very special ingredient in this process because carbon atoms have the ability to produce organic molecules that are hinged, giving these molecules the ability to create incredibly dense and completely three-dimensional structures. The 3D molecules we are describing have tremendous advantages over equivalent 2D molecules by virtue of their extremely high packing densities.

Within the first half-billion years of the earth's existence, life appeared. Life in all of its diversity has grown and flourished on our earth ever since. The English naturalist Charles Darwin[88] was among the first to recognize that all life on earth is related. Darwin's greatest insight was his realization that life is constantly changing, transforming, and evolving in response to a changing environment. In Darwin's view, all life, from the simplest

---

[88] Charles Darwin (1809–1882) was an English naturalist. His work *Origin of Species* is one of the most important works of science ever produced. Darwin developed his theories of natural selection and survival of the fittest as a direct result of his observations during his round-the-world trip on the *HMS Beagle*.

plant all the way to human beings, shares remarkably similar traits, and all life's development proceeds along very similar evolutionary paths. It was Darwin's idea that by going far enough back in the history of any two species, one could always find their common ancestor. If this is true, then all of life must be related and all living things must share many common characteristics.

Mutations and Natural Selection

There is an area in Darwin's theory of evolution and natural selection that deserves a much closer look at this point. The area in question concerns the mutations that act as a driving force behind changing life. Mutations allow natural selection to proceed, and in one sense, mutations drive natural selection. It is only by the driving force of these genetic mutations that life-forms are capable of adapting to the pressures of environmental change. Only by making a small series of changes caused by these mutations can life successfully adapt to a changing environment.

But what is the cause of these mutations, and why are they so often capable of providing successful adaptations to seemingly unanticipated environmental changes? In other words, how is it that spontaneous changes (mutations) often arrive at *just the right time* to help species make exactly the right adaptations to their changing world? How likely is it that some purely random (chance) mutations could ever hope to produce just the right changes in a specie's genetic coding that is required to facilitate the perfect adaptation to a rapidly changing environment? Is it reasonable for us to accept a hypothesis that random, unplanned, and uncaused changes are capable of serving up just the right genetic code at just the right time? Both our religious and scientific sides feel that such a hypothesis is a serious stretch. Alternative causes need to be considered. More importantly, God's role in the generation of these genetic mutations needs to be seriously addressed in light of everything that has come before within this work.

Let us now spend a few moments reviewing how nature encodes the genetic information of life. The field of molecular biology has achieved the understanding that all of the genetic coding of life on our earth is the direct result of just four different and totally unique chemical cross-links that connect the two spiraling chemical spines of the DNA molecule. Each cross-link is composed of two bases (there are four bases in all: A for adenine, C for cytosine, G for guanine, and T for thymine) that are linked at their centers by a hydrogen bond and combine into four base pairs called AT, TA, CG, and GC.

The genetic information contained in these four base pairs functions much like the information contained in a Morse code telegraph message. However, in the case of the DNA molecule, instead of dots and dashes spread out in time carrying a telegraph message, the genetic code is spread out in space along the spine of the DNA molecule. In fact, the amazing parallel between the genetic information of a DNA molecule and Morse coding is quite striking. Like the genetic code of DNA, the Morse characters that make up the alphabet contain a sequence of just four dots and/or dashes. Since genetic codes are built from four base pair choices, each set of four base pairs, like a Morse letter, must contain two bits of information. (Recall from chapter 4 that a single bit of information represents making a choice between two alternatives. It follows that a choice between four alternatives—as in DNA and Morse code—contains the mathematical equivalent of two bits of information.) See the illustration on page 108 that graphically represents how genetic information is stored within the DNA molecule.

A mutation in an individual's DNA occurs when one (or more) of the base pairs within its DNA chain spontaneously changes into one of the three alternative base pairs. That is to say, if an AT base pair were to spontaneously change into a CG base pair, a mutation would occur. Such a mutation might occur as a single base pair change, representing a small but significant modification to a gene. (Genes are self-contained sections of the overall DNA molecule chain that control certain functions related to cell growth and protein production.) A more significant mutation might affect a greater number of base pairs, creating more profound genetic changes.

FIGURE 4:
**Genetic Information Encoded
upon the Double Helix of a
DNA Molecule**

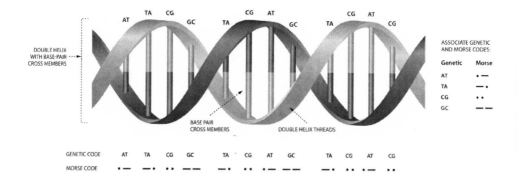

FIGURE 4:
ILLUSTRATION SHOWING HOW GENETIC INFORMATION
IS STORED WITHIN THE DNA MOLECULE

But what causes these spontaneous base pair changes in the first place? Right now molecular biology is struggling to answer this question. At this point what seems likely is that many of the spontaneous changes (mutations) along the DNA molecule are quantum mechanical in origin. A wonderful book entitled *Quantum Evolution* by Johnjoe McFadden[89] explores this subject in great depth and with much wisdom. We further conclude, based on all that has been said before, that quantum uncertainty at the moment of wave function collapse and its inherent potential for conveying God's will during *any* quantum interaction is the *primary* cause of these spontaneous mutations.

From its earliest days, the field of molecular biology (beginning with James Watson[90], Francis Crick[91], and Rosalind Franklin's[92] explanation of the structure of the DNA molecule) has recognized, as Erwin Schrödinger[93]

---

[89] Johnjoe McFadden (born 1956) is an Irish molecular biologist who feels that the source of evolutionary mutations is found in quantum mechanics. This is described in his book *Quantum Evolution*.

[90] James Watson (born 1928) is an American molecular biologist and the codiscoverer of the structure of DNA.

[91] Frances Crick (1916–2004) was an English molecular biologist and the codiscoverer of the structure of DNA.

[92] Rosalind Franklin (1920–1958) English biophysicist and X-ray crystallographer. Franklin made outstanding contributions to the understanding of the molecular structure of DNA, RNA, various viruses, coal, and graphite. She is best known for her work on DNA. While working at King's College London, Franklin made X-ray diffraction images of DNA molecules which lead directly to the discovery of the double helix structure of DNA. These images were shown to James Watson and Francis Crick, without Franklin's approval or knowledge. According to Francis Crick, it was this very data, and Franklin's interpretation of the data, that Watson and Crick used in developing their 1953 structure for the DNA molecule. Watson and Crick, who later received the Nobel Prize for their work, published their DNA model in *Nature*, and only hinted at Franklin's contributions. Franklin's contributions to the discovery of the structure of DNA have been largely overlooked by scientific community.

[93] Erwin Schrödinger (1887–1961) was an Austrian physicist and a founding father of quantum mechanics. His wave equation, which bears his name, is widely used to calculate quantum wave functions. Schrödinger also thought deeply and profoundly on the meaning and mechanisms of life. His little book *What Is Life*

predicted in 1940, that all genetic information *must* be stored within a quantum mechanical system. Mutations in a genetic information storage system occur when events transpire that change one or more base pairs from their initial choice (either AT, TA, CG, or GC) into some new choice (out of the three remaining choices). Using the Morse code analogy as an example, consider a mutation that changes a Morse H (••••) to a Morse B (–•••). This change is accomplished simply and easily by changing the first Morse symbol from a dot (•) to a dash (–)!

Examples of the possible causes of such mutations are X-rays, gamma and cosmic rays, and quantum mechanical tunneling. Tunneling is a quantum mechanical phenomenon that occurs when a particle's wave function has a nonzero probability of penetrating an energy barrier that would be expected (from classical physics) to prevent any and all penetration. Since all DNA base pairs are linked together by an interlocking central hydrogen bond, it might be possible for quantum tunneling phenomena to cause the hydrogen bonds in adjacent base pairs to break, causing, for example, an AT base pair to spontaneously change into a CG base pair. The change of only one base pair has the potential to modify a gene containing this base pair, thereby causing the mutation!

At this point, God enters the field of human awareness. Since the origin of any mutation is very likely to be quantum mechanical in nature, as we have discussed previously, God enters the picture at the very moment of wave function collapse, determining the outcome of the mutation. What we are saying is that it is God alone who determines whether an AT base pair changes into a CG or into a GC during a mutation. Perhaps God has reserved the right to decide this mutation process, and by driving the process, God's will is able to nudge along the evolution of this species in some particular direction. By sprinkling God's divine mutations among the species, God provides each species the opportunity to change along certain predetermined lines of modification.

Of course, Darwin's principles of natural selection and survival of the fittest take over at this point to determine which of the divine mutations will survive and prosper and which ones won't. Part of the wonder of this process is that God's will, the will of individual life, and environmental changes are all partners in driving the evolution of life forward. Over

---

gave birth to the science of molecular biology. He received the Nobel Prize in physics in 1933.

many generations, some mutations may prove unsuccessful in the face of unpredictable environmental conditions.

The process we are describing, which is so similar to the biblical parable of the sowing of the seeds (Luke 8:5), is one of a divine action followed by Darwin's natural selection. Clearly this process freely allows the hand of God to guide the changing forms of all life. However, it is very important to understand that the hand of God is only guiding and not determining. The final outcome of evolutionary change is determined by a two-way partnership between God's guiding hand and choices that are made by the species itself as its members cope with the challenges of a changing environment. This partnership for determining genetic change reminds us of the free-will decisions all humans must face when confronting important life choices.

Perhaps human beings were created by this ongoing process of evolutionary change that has been played out over millions of years by our animal ancestors. Perhaps God, with a wisdom we cannot possibly imagine, refined our earliest ancestors through such a process until their descendants (us), with our own particular mix of emotions, intelligence, tool-making ability (i.e., technology), and openness to godly virtue, have become a reality. Surely our human presence on earth must be a part of God's plan; perhaps it is a part of God's plan in ways that we cannot possibly imagine at this point in our history.

Life and Erwin Schrödinger

In 1927, Erwin Schrödinger, the founder of the wave mechanics mathematical formalism of quantum mechanics, was called to the University of Berlin as Max Planck's successor. When in 1933 Adolf Hitler assumed power in Germany, Schrödinger (who was an Austrian citizen) resigned his post in Berlin and fled Germany. After a period of academic roaming, Schrödinger accepted a professorship at the University of Graz in Austria. During this time Schrödinger publically stated that he could no longer live in a country (Germany) that denied the rights of its Jewish citizens (in particular the Jewish scientists who were fleeing Germany daily).

However in 1938 Austria was annexed by Germany, and the Schrödingers were once again forced to flee the country to escape the

Nazis. Their first escape was to Rome, where he contacted the Secretary of the League of Nations about his plight. From Rome, the Schrödingers went on to Geneva and Zurich and later to Oxford, and finally to Ireland, where Schrödinger had accepted a call to head up the physics department at the newly founded Institute of Advanced studies in Dublin.

Throughout this flight, Schrödinger's mind was preoccupied with the question "What is life?" This question was to be the theme of his first public lectures at the Institute of Advanced Studies in Dublin. The content of these lectures, published in book form in 1944, constituted an important breakthrough in the perception of the molecular organization of all life-forms on earth. This breakthrough paved the way for the discovery of the structure of the DNA molecule. Throughout his odyssey, one would have expected the man's mind to be consumed with fear. Instead his mind marveled at the nature of life

Although Schrödinger's "What Is Life?" lectures predated the Watson-Crick-Franklin discovery of DNA by over ten years, it accurately anticipated the implications of the Watson-Crick-Franklin discovery. In the coming decades, almost every scientist who entered the emerging field of molecular biology was first inspired to do so by reading Schrodinger's little book, What Is Life. We are sure Schrödinger never intended to found a new branch of biology, but that is the way it turned out. The history of science clearly recognizes that it was Schrödinger's What Is Life coupled with the Watson-Crick-Franklin DNA model that launched what was to become molecular biology! A valuable lesson was learned; the influence of quantum mechanics may extend much further than what its founding fathers could ever have imagined.

The main concept Schrödinger addressed in *What Is Life?* is the search for a physical mechanism that could account for the storage, replications, and transmission of genetic information down through countless generations (perhaps over millions of years). Schrödinger, with his characteristically great intuitive insight, explained in these lectures how all material things, living or not living, must obey the classical laws of thermodynamics. For living things, these laws are applied at high temperatures (the temperature of life) and concern various many-bodied systems such as gases, liquids, and most solids. Such thermodynamic systems have far too much statistical variation in their physical parameters to store the genetic information contained in living genes and chromosomes for more than a few days, let alone over a lifetime or over many lifetimes.

In Schrödinger's learned opinion, the genetic information of inheritance could only be retained sufficiently long for evolution to occur by means of an array of quantum mechanical states that could be counted on to remain unchanged indefinitely, provided they were not being subjected to some form of measurement. Schrödinger did not speculate on the exact formulation of such a quantum mechanical array; he only spoke to the general physical principles that were at work. A detailed understanding of a real quantum mechanical array containing the genetic information of life would not come about for another decade until the 1953 discovery of the DNA molecule's structure by Watson, Crick, and Franklin.

It is an interesting fact that physicists and/or mathematicians have made many of the great advances in biology when they applied their strong mathematical and physical backgrounds to the problems of biology. Gregor Mendel, whose formal training was in mathematics and physics, was so successful because he applied mathematical models to his meticulous studies of individual pea plants and their descendants over many generations. It was Mendel's mathematical exactness that made it possible for him to ultimately enhance our knowledge of the genetics of heredity and Inherited traits.

Linus Pauling's background in physics was very useful to him in his careful study of the nature of the hydrogen bonds within the base pairs of the DNA molecule. Pauling's understanding of these hydrogen bonds (which tie together all four base pairs within DNA) made it possible to understand the process of copying DNA molecules. Biology seems to be best understood when it is viewed as emerging from its underpinnings within physics and chemistry.

The work of Schrödenger, Watson, Crick, Franklin and Pauling all explain how fundamentally the biological sciences are grounded in quantum mechanics! This means everything we have learned about quantum mechanics and its role as God's mouthpiece has a direct and most profound application to the understanding the living things of our earth, including ourselves.

Charles Darwin and the Voyage of the *HMS Beagle*

We now turn our attention to Charles Darwin's theory of evolution and natural selection. It is important to understand that by its very nature,

Darwin's theory of evolution is limited to a high-level understanding of the behaviors of large populations that have existed for many generations. Because Darwin lived and worked in the nineteenth century, he had no knowledge of the physics-based molecular biology of the kind we have been discussing.

Given the state of science at the time of Darwin's work, his theory of evolution and natural selection was truly remarkable. The leap of intuition that Darwin made in developing his theory of evolution has perhaps no peer within the history of science. Evolution of species stands as one of science's greatest moves forward. However, given all of the advances in physics and chemistry since Darwin's time, we must reasonably expect that Darwin's high-level theory is incomplete and possibly inaccurate in some of its details.

Many religious people still regard Darwin's theory as a kind of battleground because they cannot see the hand of God working within evolution. However, by adding an understanding of twentieth-century physics (i.e., quantum mechanics), it now becomes possible for us to envision God's actions within Darwin's theory.

Charles Darwin was born in England in the early 1800s. His paternal grandfather was Erasmus Darwin, a famous doctor and scientist. His maternal grandfather was Josiah Wedgwood of Wedgwood china fame and a very rich man. Charles was initially sent to the University of Edinburgh to study medicine in hopes that he would follow in his grandfather Erasmus's footsteps. Unfortunately, Charles was very uncomfortable with the sight of blood, and the nonsedated surgery of his day was more than he could stomach. Charles dropped out of medical school.

His father next sent him to Cambridge to study divinity in the hopes that he might settle into the comfortable life of an Anglican country vicar. Charles completed his divinity studies at Cambridge, but his real passion at Cambridge was the few science courses he was able to take, plus the extensive fieldwork he conducted as an amateur naturalist.

In Darwin's day, many English gentlemen pursued their passion for science while remaining amateurs. Many of these amateur scientists were country clergymen, and much of their scientific work was really good. It was not unusual given the time and place to expect Darwin to follow a similar path later in life.

However, a big turning point in Darwin's life came shortly before his ordination to the priesthood. A royal navy survey ship, the HMS *Beagle*, was being prepared for a round-the-world cruise. The unique mission of the HMS *Beagle* was to conduct a scientific survey of the entire natural world, investigating all of the far-flung corners of the globe. Many round-the-world cruises had already been made by the explorers of the past, but all of these expeditions had the intention of either claiming territory for their home country or searching for profitable resources to be exploited. Never before had a circumnavigation been undertaken solely for the purpose of learning about and recording the details of the natural world in which we all live.

The *Beagle*'s master, Captain Fitzroy, wanted the ship's chief naturalist to share his cabin and of course be a gentleman of his own social rank. A suitable naturalist was chosen. However, in one of those lucky accidents that change the course of history, the chosen naturalist had to bow out at the last minute because of family concerns. The *Beagle*'s circumnavigation was expected to take at least five years, which was a long time for a man with family responsibilities to be away from home. Therefore the post was quickly offered to a young divinity student and amateur naturalist named Charles Darwin so as not to delay the *Beagle*'s sailing date

Darwin was thrilled by this turn of events, but his family was less excited than he. However, he did manage to do a successful job of selling his father on the idea of the voyage (with some help from his maternal grandfather), and after several difficult discussions, he obtained his father's permission to sail. In spite of the bad winter weather in the English Channel, the little HMS *Beagle* set sail shortly before Christmas with its new, and very excited, naturalist on board. Darwin was off to see the world. His first ominous discovery was a strong tendency toward sea sickness, which did not leave him until the *Beagle* entered the relatively calm tropical waters off the east coast of South America.

The *Beagle*'s primary mission was to map all of the territories it visited. The *Beagle*'s secondary missions were to record details of land topography, explore the local geology, and study life-forms of all kinds at every stop. This broad mission statement outlined a vast enterprise. Darwin, as the expedition's chief naturalist, was called upon to be, all at once, a mapmaker, a surveyor, a geologist, a botanist, and a biologist. This was a lot to ask of anyone so young, even in the 1830s, and Darwin was in reality only an amateur naturalist with very little formal training in any of these fields!

However, the life of a scientist was different in the 1830s, and many of the scientists in that day were of necessity renaissance men and women.

Darwin had to quickly come up to speed on such specialties as surveying and geology. As it turned out, this was very fortunate for both Darwin and all of science because his knowledge of geology and topography became of immense value to him later as he began to mentally connect the dots while collecting and inspecting fossils during his formulation of his theory of natural selection.

Darwin loved South America. He rode the pampas like an Argentine cowboy. He climbed the mountains of Patagonia. He explored the jungles of the Amazon River Basin. The sheer vastness of the species of animals and plants in South America was almost overwhelming to him when compared to the relatively few species he had been exposed to while growing up in Great Britain. Perhaps the most interesting plants and animals Darwin encountered in South America were no longer living but in fact were fossils.

This is where his new knowledge of geology came in handy because the leading geologists of the day had developed techniques for successfully dating various strata of rocks based on when and how they were created. Igneous, sedimentary, and metamorphic rocks all had their own unique creation stories. The story of their history involved specific time sequences that were necessarily based on their respective creation processes. Since the rocks themselves could be dated based on their type and on the location where they were found, it followed that any fossil found trapped within these rocks was also dateable to the same period.

Darwin now had a way of dating the fossils that were coming into his hands. He marveled at how ancient some of these fossils were. In some cases, fossils seemed to go back millions of years. Darwin marveled at how many of the fossils appeared to be similar in appearance to living plants and animals from the location where the fossils had been found. Sometimes he would find fossil animals or plants that had no living descendents, but more often the fossil creatures closely resembled the animals and plants of Darwin's own day. However, the modern creatures were always changed in some very essential ways.

Surely at this point this evidence began to send red flags up in Darwin's mind because, as he surely knew from his university religious training, Genesis 1 tells us that God created all of earth's plants and animals within a relatively short time period. Also, all of the plants and animals God created are still in existence today. Darwin now had seen the proof with his own eyes that the Genesis 1 account was wrong on both counts. The moment of creation must have occurred far back in geological time (tens if not hundreds of millions of years ago or longer).

Also, fossil evidence demonstrated beyond a shadow of a doubt that species from the past appear to be somewhat different from similar species of today in many significant ways. However, past and present species are clearly related in most essential ways. It was as if the species of the past had somehow evolved into today's species by a very gradual process that required long geological time scales to play out.

Also, the fossils from the past in a given geographical area never yielded evidence of the existence of animals or plants whose contemporary species now live in some faraway place. The fossil evidence clearly showed that all living things have undergone a process of gradual modification in certain essential ways over geological time scales—a process that produced the contemporary species we see today. A part of the mystery that took Darwin somewhat more time to unravel concerned the forces that drive this process of change in the first place. Why does all of life seem to modify itself and change in a myriad of unpredictable and unexpected ways?

When the HMS *Beagle* reached the Galapagos Islands in the Pacific Ocean west of Peru, Darwin began to gather evidence that would provide answers to his question. The answers were ultimately provided by several species of little birds in the Galapagos Islands called finches. These finches had beaks that differed slightly from island to island. But what could have possibly caused the difference in their beaks? The Galapagos were too far from South America for there to be back-and-forth finch traffic to the mainland, so the cause of the differences must have occurred right on their own respective islands.

Darwin hiked around several of the islands collecting samples and observing the finches being finches. He closely watched how the finches ate fruit and nuts from the bushes and trees growing on the island's steep volcanic hillsides. The finches seemed to be perfectly adapted to eating from these bushes because their claws and beaks were exactly the right size and shape to extract food with maximum efficiency. It seemed to Darwin that it was as if each species of finch had been designed to be perfectly adapted to the exact details of the plants, soil, and weather conditions that existed on their home island. Finches on the next island, whose plants, soil, and weather conditions were somewhat different, were likewise perfectly designed for life on their own home island. But what created the difference in design, and why did each design perfectly adapt the finches for life on their own home island?

Perhaps it was in the Galapagos that an answer came to Darwin, or perhaps it was on the long voyage back to England, or perhaps it was sometime later when he was back in England. However, sometime between 1835 and 1845, the light came on in Darwin's head and he realized that instead of his finches being designed for the specific environment of their home island, they had been self-modified in response to the specific environment of their home island. Darwin realized this process of species modification could take place in a perfectly natural way, requiring no plan and no designer.

The most accurate way to describe Darwin's idea was that he realized that over long geological time spans, the little finches themselves had somehow adapted to the changed environments on their own home islands. The finches' adaptations were slightly different on each island because the environment they encountered was slightly different from island to island. To generalize, Darwin came to realize that all organisms that have ever lived on earth have experienced what the finches of the Galapagos Islands have experienced. Even today Darwin's critical insight that environmental change itself that is the driving force behind the evolution of life stands out as one of the most profound scientific understandings that has ever been achieved by a human being.

At the end of the HMS *Beagle's* five-year circumnavigation, Darwin returned to England with the little *Beagle's* hold literally stuffed to overflowing with the flora and fauna (and fossil records) of the entire world. Captain Fitzroy thought Darwin was a little daft because he endangered the ship with the weight of his vast below-decks collection. But Darwin's collection was the evidence he would need to support what was to become his theory of the evolution of species.

In spite of his seminary education, Darwin never chose to be ordained. Soon after his return to England, Darwin was married. Charles and his wife, Emma, found a country house to settle down in, raised a family, and were very happy with their lives. As an heir to the Wedgwood china fortune, Charles had no concerns about money. He soon settled into the lifelong role of a self-styled naturalist operating within the domain of his new country home. He was a doting father to the couple's five children and by all accounts a good husband. He never sought an academic post and rarely published. Most of his scientific contributions came in the form of letters to friends describing his latest findings, insights, and hypotheses. Charles Darwin was a quiet, soft-spoken man who avoided controversy, traveled very little, and relished his life as a parent and a county gentleman.

It was these very components of Darwin's personality that may explain the long time gap between the end of his world circumnavigation and his publication of the theory of evolution—a period of nearly thirty years! Very early on, perhaps when he was still on the *Beagle*, Darwin must have realized that the theory of evolution he was still formulating would be controversial at best and perhaps unpalatably revolutionary to the conventionally religious world of Victorian England. It may have simply been his desire to avoid the unpleasant consequences of being the bearer of revolutionary ideas that restrained Darwin from publishing any sooner. ("A man who speaks the truth must have one foot in the stirrup," as an old saying goes.)

However, many of Darwin's friends knew what he was up to. When he did finally publish, it was only after the constant pressure of his friends telling him that others were working on similar ideas and if he didn't publish soon, he would risk losing his priority. In this respect, Darwin stands in the proud and grand tradition of Copernicus and Newton, who both had to be pushed and cajoled by their friends into publishing their greatest scientific works. Perhaps all three of these giants of science were reluctant to publish what was to become their greatest work out of both a fear of controversy and an uncertainty over the correctness of certain parts of their theories. They need not have worried. In all three cases, their theories developed a life of their own, carrying the initial ideas into a continuing process of refinement driven by countless soul—and fact-searching reviews by their fellow scientists.

Upon his return to England, Charles still had some difficult conceptual issues with his theory that needed to be faced up to and overcome. The first issue was the question of what mechanism was responsible for slowly adapting life to a changing environment. The second problem was concerned with how the changes in a species are captured and recorded so they can be passed along to their descendants. The answer to both questions lay with the causes and effects of mutations.

Mutations are the spontaneous changes that occur in an individual that distinguish it slightly from its fellows of the same species. Remember that Darwin had absolutely no knowledge of DNA or molecular biology in general. It would be about one hundred years until Schrödinger would give his Dublin "What Is Life?" lectures claiming that a quantum mechanical process is the only possible means for storing genetic information. Darwin only knew that mutations took place in all life-forms and that some mutations survived to be passed on to later generations but most did not.

Darwin did not try to make any hypothesis about what might be the microscopic mechanisms for these mutations. It was just as well to accept them as both ubiquitous and random because it would be another hundred years before the scientific underpinning to understand the nature of mutations would become available in quantum mechanics. These very random mutations, however, are the mechanism for changing life in response to a changing environment.

Darwin's hypothesis was that mutations happen all the time and the overwhelming majority of them do not equip an individual to live any more successfully in a changed environment. However, on a few rare occasions, an individual might receive a very special mutation, making this and only this individual slightly more fit to survive within the newly changed environment. This slightly more fit individual will, on the average, live longer than his or her nonmutated fellows.

This improved fitness bestowed by the successful mutation will on the average, because of increased longevity, allow the mutated individual to mate more often and produce more offspring than his or her nonmutated fellows. Therefore, the mutated gene associated with this successful individual will spread into the species' gene pool in somewhat greater numbers than the equivalent nonmutated genes. Over time the mutated gene will spread, influencing more and more individuals. In time another successful mutation will occur within the species, and the process will repeat itself and continues. (Remember that this is a process that plays out over geological time spans, which is to say over hundreds if not thousands of generations.)

Since unsuccessful mutants do not live long enough to mate very often and pass along their genes, their unsuccessfully mutated genes are quickly removed from the gene pool. On the other hand, the genes associated with successfully mutated individuals are passed on to many individuals and in time become the nominal form of that particular gene.

Briefly, this is the slow process of evolution at work. The process is self-correcting. Unsuccessful mutations will quickly die out in the gene pool, and only successful mutations will be adopted by the species as a whole. The process naturally tends toward greatest adaptation in a changing environment, so it is a never-ending process of continual refinement because the environmental changes on earth are never-ending. Therefore, there really is a no final form to any species. All species find themselves in a constant state of change in response to the changing environment its individuals are always confronting. Predators adapt to changing prey to become more adept hunters, and the prey adapts to changing predators to make themselves more difficult targets of predation. If either predators or prey were to ultimately gain the upper hand, at least one species would quickly die out.

Darwin saw the process of evolution as the dance of life engaged in by all species, including our own. The evolutionary process is continually creating new life-forms and causing old life-forms to go extinct. We humans, as a part of this process, are closely related to the great apes, like chimpanzees. Darwin thought humans and chimpanzees shared a common ancestor who long ago had became extinct.

All of Darwin's evolutionary theory is in direct conflict with the account of creation given in Genesis 1. In the Genesis account, God created all of the animals within a short period of time, and all of God's creations possessed the same form in which they appear today. According to the Bible, God created humans in God's own image (more on this in chapter 7) and gave humans dominion over all the animals. In the Bible, a clear line of distinction is drawn between animal life and human life. Clearly the Genesis account of creation and Darwin's evolutionary theories are locked in head to head conflict on this point.

Surely Darwin could see this conflict coming, and perhaps its unpleasant consequences were among the reasons Darwin was slow to publish his work. In the end, Darwin's friends convinced him that if he didn't publish soon, he would surely be beaten to the punch and ultimately lose his priority. He finally agreed, and over a two-year period, Darwin wrote his Origin of Species. Like Newton's *Principia Mathematica*, Darwin's *Origin of Species* was only published with a lot of help from the author's friends.

Once Darwin's book was out, it created a very predictable controversy. The church was up in arms, as were many laypeople. Evolution simply did not suit the Victorian sensibilities of many people. But for other, more open-minded people, evolution made a lot of sense.

Darwin got at least as much support for his theory as he got criticism. However, because of the nature of his personality, Charles personally shied away from the inevitable public confrontations. This task was taken over by his friend T. H. Huxley. Huxley, who became known as "Darwin's bulldog," gladly debated any and all comers.

At a meeting of the Royal Society, Huxley debated a prominent Church of England bishop concerning the conflicts between the Bible and evolution. This bishop's nickname was "Soapy" because he had the reputation that whenever he got into hot water, he came out of it smelling like a rose. However, on this occasion the good bishop, who strongly protested that there were no monkeys in his family background, ended up sounding a lot like a monkey when he debated Huxley. Unfortunately, it is this kind of debate that has increased the gulf between science and religion.

As unbelievable as it may sound, the Bible versus evolution debate continues unabated to this day (largely in America and less so in Europe). Battle lines are drawn, and no one is giving any ground. Positions on both sides have hardened. Regrettably, politics have gotten into the act, and who knows where truth is anymore? Sadly, it is an old story.

The final act of this saga was written in 1953 by a young American biologist working in England, an English solid-state physicist, and an English X-ray crystalographer who combined forces to give form to Schrödinger's concept of a quantum mechanical basis for the storage of genetic information. Their work would change the world of biology forever. Quite simply put, James Watson, Francis Crick, and Rosalind Franklin discovered the form of the giant organic molecular DNA and with it the source of all heredity and the mechanism for evolution acting at the molecular level.

The DNA molecule, which may also be called an a-periodic crystal, was envisioned by Watson, Crick, and Franklin as a kind of molecular ladder that is twisted in such a way that its upright portions form a pair of helixes and its cross members (called base pairs), which link the double-helix together, are (as we have seen) the coded elements with four distinct flavors. The base pairs and their form can be thought of as functionally similar to, as we have already seen, the dots and dashes of Morse code. A gene is comprised of a sequence of base pairs, and a chromosome is composed of a sequence of genes.

It should be noted that not all base pairs in an individual's DNA are active participants in creating the genes and chromosomes of life. Many of the base pairs in a genome are simply placeholders, perhaps left over from the DNA of a long-extinct ancestor. The human genome is a mapping of each and every base pair in all of the gene sequences that make up every chromosome that comprises human DNA.

The Watson-Crick-Franklin DNA model provides a way of understanding the mechanisms of mutation. Since mutations hold a pivotal role in Darwin's theory of evolution, it is extremely enlightening to gain insight into their physical origins. Each base pair in a strand of DNA contains a small piece of the genetic coding that determines the ultimate form of the individual. Since there are four choices for each base pair, if anything happens to the individual's DNA causing, say, an AT base pair to become a TA base pair, it is equivalent to rewriting one letter of the individual's genetic coding. Perhaps rewriting just one letter (one base pair) makes a significant difference, corresponding to one of Darwin's mutations that, if successful, could drive the evolutionary process forward.

What physical causes could change an AT base pair into TA base pair? There are several causes of mutations, including exposure to ionizing radiation, exposure to certain chemicals, exposure to ultraviolet light, and quantum tunneling. Let us now explore the possibility of quantum tunneling in more detail. In fact, all of the causes of genetic mutations are quantum mechanical in nature, but quantum tunneling is the cause that is most easily understood in light of everything we have discussed thus far.

Quantum Tunneling

We now will focus on how mutations could be caused by quantum tunneling. Tunneling is a completely quantum mechanical phenomenon in which a particle is able to tunnel through an energy barrier in a ghostly way that is not allowed by classical physics. The base pairs of the DNA molecule are held together by a hydrogen bond located near the center of the base pairs. (Base pairs are the molecular cross members within the DNA double helix.) By disturbing these hydrogen bonds, one side of the base pair may be modified, changing it, for instance, from an AT base pair to a TA base pair.

Of course, since tunneling is a quantum mechanical phenomenon, it is subject to all of the same quantum strangeness as any other quantum phenomenon (i.e., uncertainty, coherence, probability, and superposition). Therefore, exactly when and where a base pair will spontaneously change from a AT to a TA, by virtue of tunneling, is not only unknown, but it is also unknowable because of the uncertainty principle.

This means that if God ultimately controls the when and the where of individual quantum events, then it is God who holds an option on when and where a particular mutation will occur. This means that God always holds an option on determining what mutations will drive forward the process of evolution!

Note that God has not violated any of the laws of science by exercising this option. "Do not think that I came to destroy the law or the prophets. I did not come to destroy but to fulfill" (Matt. 5:17). God, in determining the outcome of a mutation, has not violated the principle of conservation of energy or any other scientific law, for that matter. God has committed no violation of causality because the ultimate cause of these mutations is inherently lost to us in quantum uncertainty. Therefore nothing has happened in the God-determined mutation process that might threaten the truth of science. However, God's presence in this critical process also strengthens religious truth, and herein lies the paradigm shift. Everyone wins!

At the same time, a new element has been added to the story of evolution with the suggestion that God and life itself are in fact cocreators of new life. We might think of God as the sower of the seeds we call mutations. When these seeds produce successful mutations according to God's plan, they are further tested for rightness within in a changing environment by the process of natural selection as described by Darwin. This testing for environmental rightness by natural selection is, in our opinion the first appearance of the gift of free will, which is discussed in the next chapter.

God offers the gift of new living possibilities, but it is life itself that ultimately determines whether to accept or reject God's gift. We personally find it far more reasonable to believe that God has personally chosen the quantum seeds of successful mutations rather than believe these successful seeds could ever have appeared out of a completely random and meaningless process of unintended chance. This process clearly represents a paradigm shift on the part of both science and religion. Religion is asked to see God's creations not as simple one-shot occurrences but as ongoing unfoldments. Science for its part is asked to accept God's pivotal role within this process as God grants the gift of creative newness to key individuals.

We think it is a part of God's plan for life on planet earth for God to personally guide life's evolutionary process through a long series of tiny suggestions called mutations (each occurring clothed in quantum mystery), making these suggestions appear indistinguishable from the more visible process of natural selection. We are reminded of how the Wizard of Oz,[94] in the movie of the same name, hid behind a curtain to avoid recognition.

As we have said before, God in this role reminds us of the God of Genesis 2, who is described as a gardener, doing what all gardeners do: planting seeds, nurturing the young plants, and at the right time, gathering in the harvest. Perhaps mutations are the seeds that are spread by God as he exercises his option to make the final choice between quantum alternatives. Life itself is the young plant, and we humans, with our great intellectual powers and spiritual connections to God, are the harvest (but perhaps not the only harvest, as discussed in chapter 8).

Jesus spoke in many of his parables about planting seeds (Luke 8:4-8). He talked about how some seeds grow and others do not and some plants are gathered into the harvest while others are choked by weeds. It is our feeling that the God of Genesis 2 and the God of Jesus's parables exists in perfect harmony with God as the master sculptor of life. God in this role drives life ever so subtly this way or that way in accordance with God's plan.

Coherence

For a biological system to experience quantum mechanical effects (like the tunneling process in mutating DNA and the firing synapses in the nerve tissue of the brain), it is necessary for the so-called quantum coherence distances within living tissue to be long enough for these effects to occur. Some examples are that base pair lengths within mutating DNA must not be short when compared to the coherence distance and that the length of a brain synapse must not be short compared to the coherence length for quantum effects to occur. But what is coherence distance, and what does it mean?

---

[94] *The Wizard of Oz* is a children's story by L. Frank Baum.

Among all natural phenomena, waves are unique in that they exhibit certain behaviors called constructive and destructive interference. The potential for two or more waves to experience constructive and destructive interference is called coherence. Although in this work we are primarily interested in quantum phenomena, all types of wave phenomena universally experience coherence. Normal, everyday physical waves such as sound waves, radio waves, or the seismic waves associated with earthquakes propagate with definite amplitude at a definite frequency. However, it is a universal wave property called phase that ultimately determines if a wave experiences constructive or destructive interference.

Phase is associated with the relative alignment of two or more waves. If two waves are perfectly aligned, they are said to have zero relative phase shift and will experience constructive interference if they interact. Constructive interference means the resulting wave will have two times the amplitude of the individual waves. If the two waves are completely out of alignment (i.e., the peak amplitude of one wave corresponds to the valley amplitude of the other wave), they are said to have 180 degrees of relative phase shift, and if they interact, these two waves will experience destructive interference, causing them to literally cancel each other out! Quite literally, nothing is left. Every other relative phase shift will result in an interaction midway between these two extremes.

An example of constructive interference is in the whispering galleries of St. Paul's Cathedral in London, where listeners standing in the elevated gallery on one side of the cathedral's dome can clearly hear a private, whispered conversation that is taking place on the other side of the dome. Many a juicy secret has been shared with unsuspecting strangers by way of this phenomenon.

On the other hand, an example of destructive interference is the use of special headsets to bring about a reduction in the sort of noise that is often experienced by people who are exposed to prolonged levels of high audio noise, such as passengers on airliners. In noise-reducing headsets, the background noise from jet engines is amplified by electronic circuits within the headset and shifted in phase by 180 degrees before being recombined with the sound that is reaching the user's ear. The result is almost complete cancellation of the airplane's engine noise.

In completely analogous ways to the whispering galleries and noise-canceling headsets, subatomic quantum particles, acting as waves, experience exactly the same wave interference phenomena. The situation is particularly dramatic in cases where the particle's wave function may be thought of as the superposition of two or more elementary wave functions, each one representing a possible outcome of a measurement.

A classic example of quantum interference is the famous double-slit experiment (see chapter 2) in which electrons are fired at a metal plate containing an upper and a lower slit capable of allowing electrons to pass through the metal plate and impact on a photographic plate that is located just behind the metal plate. It is the purpose of this photographic plate to record the position of each electron impact as it passes through the slits in the metal plate.

Reason would suggest that most of the electron impacts would be recorded on that portion of the photographic plate that lies just behind each of the two slits. In fact, if such an experiment were really conducted, the results would be completely unlike what we expect. The real outcome would be that most of the impacts lie at a point midway between the two slits, where reason would tells us no impacts can occur because the electrons are totally blocked from reaching the photographic plate through the solid material of the metal plate.

This reasoning is based on our expectations from everyday life and classical physics. However, quantum mechanics predicts a completely different outcome based on wave function interference phenomena. From a quantum mechanical viewpoint, electrons may be represented as a wave function composed of the superposition of two wave functions, one passing through the upper slit and a second passing through the lower slit.

Even with a single electron, both wave functions simultaneously pass through their respective slits, emerge on the photographic plate side of the metal plate, and are subject to wave interference (coherence) phenomena. If the slits in the metal plate are spaced apart by a distance corresponding to a 180-degree relative phase shift between the two wave function superpositions, the complete cancellation of one wave function superposition by the other superposition will occur directly behind each slit. No impact will be recorded on the photographic plate, even though this point lies directly behind the slit where electrons are expected to pass.

In the same way, by adjusting the spacing between the slits and the spacing between the metal plate and the photographic plate to ensure the two wave functions have exactly zero degrees of relative phase shift when they arrive at a point on the photographic plate that is exactly midway between the slits, the impact of the electron will most likely be recorded at this position by the photographic plate, even though it has seemingly passed through an impenetrable barrier represented by the metal plate!

The double-slit experiment stretches our ability to grasp the meaning of quantum mechanics because it asks us to accept the reality of the impossibility of an electron passing through the slits where it should pass and the reality of the electron passing right through the metal plate where it should not pass. Most difficult of all, we are asked to accept the fact that the electron interacts with itself when it brings about these wonders! Nobel Prize winner Richard Feynman once remarked that the double-slit experiment, more than any other quantum thought experiment, fully embodied the wave nature of quantum mechanics, exposing its completely nonsensical weirdness that is so difficult for any of us to get our heads around.

John Archibald Wheeler,[95] a student of Neils Bohr and a longtime physics professor at Princeton University, promoted the idea of a quantum particle being measured by the environment. With this statement, Wheeler meant that independent of any manmade experiment, a quantum entity (such as an electron) that comes into contact with a population of other particles (such as a gas) will sooner or later collide with one of these particles, exchanging energy and momentum in such a way that the particle's wave function collapses. At the point of collapse, a quantum particle is said to be measured by the gas particles of the environment, as surely as if it had impacted an experimenter's photographic plate and left a smudge.

A quantum particle being measured by its environment is a little like the old philosophical question, "If a tree falls in a deserted forest where no one can hear it fall, will it still make a noise?" I guess this question can be argued back and forth, but the fact remains that a particle, after the collision, has been altered by a particle-like exchange of energy and momentum and by the collapse of its wave function, which causes all quantum interference phenomena to cease.

Therefore, a concept has emerged that quantum interference phenomena can only exist over some limited distance before a measurement occurs within a population of other unrelated particles. This limited

---

[95]   John Archibald Wheeler (1911–2008) was an American physicist and student of Neils Bohr. He made many contributions to the continued understanding of the meaning of quantum mechanics.

distance is called the coherence length of the environment, because it is only within the coherence length that quantum interference interactions can occur. Coherence length is strongly dependent on temperature, because these particle-to-particle interactions are greatly enhanced at elevated temperatures.

As an example, an extremely long coherence length is found in low-temperature superconductivity. If such a conductor becomes a superconductor, the wave functions of its electrons will literally extend throughout the entire length of the conductor! In the case of superconducting electromagnets (of the kind used in MRI[96] machines), the coherence length can be many meters!

However, in the case of living things, the situation is very different than in the case of superconducting magnets. Since the metabolism of living things requires temperatures in the range of 20°C–120°C, the coherence length of quantum phenomena within living things must take into account these relatively high temperatures. Temperature needs to be taken into account when evaluating the realism of quantum phenomena in living things such as DNA molecules and brain synapses. In all likelihood, mutations within DNA molecules could be caused by quantum tunneling since the coherence length of this environment is on the order of tens of nanometers, a dimension that is much longer than base pair to base pair spacing along the axis of the DNA molecule.

However, the coherence length within a brain may not be sufficiently long to explain quantum phenomena in brain synapses. The area of biological quantum phenomena remains on the forefront of research, and it may be some time before a scientific consensus emerges concerning the full range of quantum phenomena in living things (apart from DNA molecules). For the moment the best that can be said is that these possibilities exist, and we must try to understand their implications and likelihood.

As we have seen, life in all its complexities is hard to fathom from a purely scientific point of view or from a purely religious point of view. However, by combining our scientific knowledge that spells out the details of life's genetic memory mechanisms residing in the DNA molecule and

---

[96] MRI stands for magnetic resonant imaging. It is widely used as a medical diagnostic tool. Its principle of operation depends on observing the quantum mechanical spin of protons within the hydrogen atoms of living tissue.

religious knowledge reminding us of God's intentions to create all life (including our own), we see how God's will can be perceived in all life by the evolutionary process called natural selection. We also recognize the quantum mechanical channels of godly communication that reside in each and every base pair in each and every DNA molecule.

It is through the agency of these God-inspired mutations that the seeds of new life emerge to be tested by the processes of natural selection. However, it is these godly mutations that plants seed in the process of evolution with the raw material of creative genetic ideas, some of which will be selected as additions to the gene pool as these new variations on old life go through the crucible called natural selection and survival of the fittest. Here we meet face to face the combined truths of science and religion supporting each other all along the way in creating a whole and undivided truth.

The Quantum Brain

In our search for a moral compass, each of us pray that God will grant us the knowledge of God's virtues. But how does God grant a prayer for virtue? A virtue must be grafted into our human emotional structure if it is to become a free-will alternative to the emotions of our survival instincts (such as anger, fear, and lust).

Consider first the workings of the human nervous system. In human bodies, this system is made up of the brain, the spinal column, and all of the nerves of the body. Inside the human brain, there are truly an astronomical number of brain cells, which are called neurons. These cells roughly play the same role as logic gates do in a digital computer. However, unlike a computer's electronic logic gates, which have at most five or six interconnections to other gates, the neurons of the human brain are interconnected to one another in a great number (perhaps hundreds) of highly complicated ways.

These interconnections are along axonal pathways called dendrites that are turned on and off by switch-like elements called synapses. Each synapse is a kind of electrochemical switch that allows information to flow (or not to flow) along the axon pathway connecting the neurons. The neurons comprise the gray matter of our brain, while the axons, with their white insulating coatings, are the long, interconnecting fibers that comprise the brain's white matter.

Human brains are about 50 percent gray matter and 50 percent white matter. Our ability to remember events and cognate rational thoughts is linked to the operation of the neurons and their synapses. There are a vast number of synapses in the human brain (perhaps one followed by eleven zeros). This astronomical number of synapses is on the order of the number of galaxies in our entire universe (exceeding 100 billion). Based on these numbers, the potential for our brains to receive and process information is extremely high.

Each human brain contains two types of synapses: chemical synapses and electrical synapses. In a chemical synapse, information is passed across the synapse by the movement of chemical ions inside the synapse's structure. In an electrical synapse, it is the movement of electrons across the synapse's structure that transmits the information. The size of an electrical synapse is about one-tenth the size of a chemical synapse and is on the order of one nanometer (a nanometer is one meter divided by one followed by nine zeros). For such small dimensions, it is entirely possible that quantum mechanical effects are significant factors in controlling the information that passes across the synapse.

Ongoing research in this area is focused on providing definitive answers to this question. To be certain quantum effects are involved within the brain's synapses, it is first necessary to determine over what distance a quantum wave function can maintain coherence within the environment of the human brain.

The coherence distance within the human brain is determined by thermal wavelength, which is a measure of how far an electron's wave function can extend before the electrons it represents are forced to interact with the environment (which is quite warm in the case of the human brain, leading to robust interactions). However, the thermal wavelength is short at high temperatures, which means brain synapses must be small to support quantum interactions. In superconductors that operate at temperatures close to absolute zero, the electron's wave function may remain coherent over many feet!

When an interaction with the environment occurs, what happens in effect is that a quantum measurement occurs, destroying wave coherence (which is to say collapsing the wave function in the language of the Copenhagen interpretation). If the synapse's gap is small enough to sustain quantum mechanical coherence (at a normal brain temperature of 98.6°F), an electron passing across the synapse will consist of a superposition of wave functions, representing an electron passing through the synapse and an electron not passing through the synapse.

The gap within a human electronic brain synapse is on the order of one nanometer, whereas the thermal wavelength of an electron at the temperature of a human brain is about 0.20 nanometers. This means quantum mechanical coherence may or may not be a significant factor in a structure the size of a human brain synapse. As in all quantum interactions, it is the wave interference effects experienced by the two wave function superpositions that determines the final outcome of the measurement.

Since the synapse is assumed to be operating as a quantum mechanical system, all that can be determined about this system is the probability that a certain outcome will occur. Our basic assumption throughout this work has been that God has the ability to know and control the exact outcome of any quantum measurement. (To slightly paraphrase Einstein, "God doesn't play dice" in any way, but God knows and controls the outcome of any quantum interaction before the dice are rolled.) Therefore, as in the case of DNA mutations, God retains an option to determine when and if a specific bit of information will flow across any one of the billions of synapses within the human brain.

This concept has enormous implications. Consider a human being in meditation or prayer. The person in meditation attempts to quiet his or her thoughts to achieve a peaceful awareness of the small voice within the heart. This is the small voice that is widely believed by religious people to be truly the voice of God. The voice of God can carry insights, provide answers to difficult questions that have been prayerfully posed, and affect a change of the emotional heart within the soul of people who have a committed prayer and meditation practice.

Likewise, a dreamer will sometimes receive dramatic insights into difficult life situations in the form of symbolic dreams. It is widely believed by many religious people that dreams are God's way of communicating with us when we are asleep. For instance, in St. Matthew's gospel, Joseph's decision making is directed by dreams. As with meditation and prayer, God's opportunity to affect our dreams is linked to God's ability to exercise control over the information that passes across our brain synapses. Since the number of synapses within our brain is on the order of the number of galaxies in our universe, God, using quantum synchronicity, has plenty of raw material to work with to paint the beautifully detailed pictures of our dreams.

Not all dreams are of great significance, and perhaps some dreams are just our mind's way of rehashing the events of the day. However, every so often there occurs what we call a big dream—a dream of such profound significance that we often wake up and immediately recall our dream down to the finest detail (recall Allen's dreams described in the preface). We find that by revisiting these big dreams from time to time, a great richness of new insights will continue to unfold for years to come. The following is a dream of Fritz's that is the kind of dream we are talking about here.

> I was an observer on a wide boulevard in an unfamiliar city. An elderly couple—homeless people, I thought—pushed a shopping cart with their belongings down a side street past shops and stores. With them and apparently related to them were ten or more children of different ages. All wore nightgown-like brown garments that I remember children wearing in Lebanon when I was there. These kids were playing around, making fun of not only the old couple but also of the merchants and passersby. I felt overcome by a wave of empathy for the old couple and also for the children. Happy though they were, they were obviously poor and needed schooling, shelter, and food.

> Next I found myself seated at the table of my teacher and his friends. There was food and drink and conversation. The room, the table, and the people around it looked like the scene depicted by Leonardo DaVinci[97] of Christ at his last supper with his friends. I took my seat at the small end of the table to the left of the Master. I did not join the conversation but sat there looking downcast, preoccupied with my thoughts.

> The Master looked over to me and said, "What's the matter with you? What makes you look so sad?"

[97] Leonardo da Vinci (1452–1519) was an Italian scientist and artist. When we think of a Renaissance man, we think of Leonardo.

I answered, "I have seen an elderly couple on a city street, tired and struggling and surrounded by children. All of them were homeless. I wanted to help them, but I knew not how."

The Master's face grew serious for a moment. Then he rose from the table smiling. The back wall of the room disappeared, exposing a green hillside to view. The Master strode up that hillside. With each step, his stature grew. Then he turned, facing us. His arms spread wide, he rocked from side to side, and at that, those brown-clad children I had seen with the old couple in the street came from left and right up the hill toward him. Holding on to him and to each other, they danced with him.

Then he said with a loud voice, "In this business you have to stretch, stretch, stretch with all you have in you to serve these children and the world!" At that the children left the scene again. He came down and seated himself at the table again. The wall built back in behind him. He still smiled and beckoned to an attendant behind him. He told him something. The man left and came back with two books in hand. The Master gave them to me and said: "Here, these are for you." I hugged the books, bowed, and left the room, awaking into objective consciousness.

I got up, took the Bible from the shelf, and opened it to Matthew 13:51. I read: "'Have you understood all of this?' 'Yes' they answered, to which he replied, 'Every scribe trained for the Kingdom of God is like a householder who can bring from his storeroom both that which is new and that which is old.'"

This was the time of my life when my wife, Stella, and I found each other at our work place. We both were grounded on the treasure trove of faith. In that faith, we did our service work together. Often enough our service endeavor has been stretched to the limit of endurance, but it stayed alive in faith, even beyond the time when death parted us, just as the dream had prophesied.

Perhaps the most significant outcome of our prayers and meditations are the changes God brings into our emotional heart. God, like an old alchemist, finds ways to change the base metal of our emotional makeup into the golden virtues of God's plan. If we give God the chance, our fears can be changed into courage, our angers into patience, our lusts into affection, our greed into generosity (remember Scrooge of Charles Dickens's story The Christmas Carol), and perhaps even our hatreds into love. (Remember, Jesus commanded us to love our enemies.)

But the choice is always ours alone. God's gift of virtue is free, and in the last analysis, it is always our own choice to indulge in the temporary pleasures of our survival emotions (gifts from our evolutionary ancestors) or to make the conscious choice to up-end our emotional glass and ask God to transform our survival emotions into the true gold of God's virtues. God understands—better than we do ourselves—that our human survival emotions are an indelible component of our physical structure. He knows that no evil is involved on our part when we experience these inherited emotions that are so much a part of us. God provides us with God's virtues so we may grow and emotionally stretch ourselves by freely choosing these godly virtues as we confront the hard choices that are required by the living of our lives.

Perhaps God's purpose is to bring about a spiritual evolution within the human race, paralleling the biological evolution of our bodies. We grow most when we freely choose God's virtues, not when we blindly obey a set of laws, the transgressions of which are to be blindly repaid with a crushing burden of guilt.

CHAPTER 7

HUMAN NATURE

The Human Quest for God

Perhaps the most human activity we are ever to engage in is our individual and collective search for God. There are no words with which we can adequately describe God. There is no way for humans to understand God. However, the human central nervous system is in fact uniquely crafted to support our search for God. The quest for God is our purpose in life and our destiny as human beings. There is no higher calling!

To speak of God, we are forced to use art, poetry, music, and metaphor. Artists often depict God by using human images arranged in distinctly superhuman ways. Consider the Buddhist and Hindu art of east Asia. In the temple statues of these cultures, God is often represented as a humanlike individual with multiple faces and/or multiple sets of arms and hands. The metaphor projected by this art seems to be a perception of God as a multiplicity of personalities. Another example would be the Trinitarian God of Christianity. In the case of God the Trinity, the multiplicity of God is three. However, for the Asian religions, God's multiplicity seems to be limitless!

Consider another model for representing God's multiplicity. This model comes out of the science of quantum mechanics, based on the Copenhagen interpretation. The measurement of a quantum mechanical system is carried out by a measurement system that aims to accurately determine certain parameters at the expense of certain other parameters. The Heisenberg uncertainty principle guarantees that not all of the parameters of a quantum mechanical system *can* be measured simultaneously (see chapter 2 for a detailed discussion of the uncertainty principle).

According to the uncertainty principle, the more accurately we measure a certain parameter, the less accurately we are *able* to measure another complementary parameter of the same system. All of the possible outcomes of a quantum mechanical system can be represented by what is called a wave function superposition. The experimental measuring equipment is configured to enable the wave function superposition to interfere with itself, thereby determining the behavior of the wave function at the actual point of measurement. At the final moment of measurement, according to the Copenhagen interpretation, the wave function is said to collapse as the information contained in the measured parameters is captured by the measurement system and irreversibly recorded. It is not possible to predict what the experiment's outcome will be before running the experiment because of the probabilistic nature of all quantum mechanical systems. Only by running the experiment can an outcome be determined. If the same experiment is run again, the outcome *will* be different because of the quantum entity's inherently probabilistic nature.

Consider now what it might mean to use quantum measurements as a model for a human interaction with God. A human being (or more specifically an individual's central nervous system) corresponds to and in fact is (on a very large scale) a quantum measurement system. The wave function corresponds to God. The wave function's superposition corresponds to God's multiplicity. An individual may be searching for answers to a difficult question, or he or she may be seeking help to understand what choices to make when facing a vexing life decision. These questions, and their intentions, represent the laboratory configuration of a quantum mechanical measurement. Since God's multiplicity is limitless, depending on what one is searching for, there could be a very large number of super-positioned faces of God interacting with each other to provide the answer to the posed question.

As an example, consider a woman who is searching for insight into whether God exists or does not exist. There are two superpositions associated with this question; either God exists or God does not exist. These are the only two possible outcomes. By prayerfully posing this question, the woman opens herself to inspirations that may help her reach a conclusion about what is true and what is not true.

Quite literally, within this woman's mind, God's existence and God's nonexistence interfere coherently with each other to produce the insight that in the final analysis will lead her in the right direction. Her answer may come in the form of an internal impression (as in a dreams), or alternatively her answer may come to her externally (perhaps in a conversation with an unexpected stranger or while she reads a book or listens to the radio). Ultimately, the outcome for her may simply be a shift of her attitude toward the question. She might reach the conclusion that in human terms, God exists and does not exist all at the same time! What we are describing here is a completely natural process that few of us ever recognize as such. The answers to the most profound of life's questions may simply show up by chance on a radio show, in a randomly chosen passage of Scripture, or in the chance words of a stranger. We don't so much grasp the truth as reach out for the truth.

The Serpent Beguiled Me and I Ate

There is no consistent attitude among religious people toward evil, with one notable exception: the Gnostic Christians of late antiquity. This is particularly strange when you consider that evil's effects are constantly visible. Just pick up a newspaper or turn on the television and you will literally be bombarded with a daily chronicling of the effects of evil on our world. Jesus made repeated references to the "prince (or principal) of this world" within the canonical gospels, implying there is a strong malevolent force at work within human existence.

In the Old Testament, only Adam and Eve's temptation by the snake and Job's testing by the devil make any direct reference to the appearance of the reality of a personified evil. Among orthodox Christians, Augustine of Hippo was the first theologian to define evil. For Augustine, evil was simply the absence of God. This concept is very hard for people of later times to accept. Thomas Aquinas elaborated on Augustine's concept of evil by redefining evil as the lack of, or the insufficiency of, God's divine order within the world. This concept is equally hard for our twenty-first-century minds to accept.

A little later in history (the thirteenth century), Dante[98] concluded that we all make our own hells by defying God's order. For Dante, hell is filled with the spirits of those who reaped what they have sown. In Dante's

---

[98] Dante Alighieri (1265–1321) was an Italian writer and philosopher. His most famous work is the *Divine Comedy*.

view, Satan, the ultimate rejecter of God's order and will, lies frozen within hell's deepest pit, immobilized by ice (Dante's hell was not a hot place). Still later, Milton explored Satan's fall in his work *Paradise Lost*. Milton's opinion seemed to be that Satan had led a revolt in heaven out of envy for God's son. When Satan lost this fight, he was banished forever to hell, a place from which he routinely visits earth to torment human beings with his constant and creative tempting.

The most creative treatment of evil seems to remain in literature and poetry (Dante and Milton are the most notable but not exclusive examples of this trend) and in the popular mythology of the time. Modern-day writers of fantasy literature (such as C. S. Lewis, J. R. R. Tolkien,[99] J. K. Rowling,[100] Gene Roddenberry,[101] and George Lucas[102]) have done a lot to explore the meaning and manifestations of evil in all its many forms. Many, if not most, people in Western culture feel that evil is a spiritual force that stands in opposition to God and everything that is good. Whatever this force is, it seems to always be at work tempting us (like Adam and Eve) to act in opposition to God's will and fall into sin. Orthodox Christians, unlike Gnostic Christians, have no well-defined boundaries between what is good and what is evil. They must struggle along with the reality of an ill-defined boundary.

Eastern religions, such as Hinduism and Buddhism, do no better in defining evil. These religions teach that each of us goes through life collecting the personal karma that is associated with our own interpersonal interactions. They are generated in particular when things go badly between ourselves and our neighbors. These karmic debts must be repaid (worked out with our creditors and debtors) in future lifetimes. It is the attachment to these karmic debts that keeps individuals returning and returning to future existences. But evil in this context does not exist as an independent external force. Evil from the eastern point of view is better understood

---

[99] J. R. R. Tolkien (1892–1973) was an English fantasy writer. He is best known for his *Lord of the Rings* trilogy.

[100] J. K. Rowling (born 1965) is a Scottish fantasy writer. She is best known for her *Harry Potter* series of books.

[101] Gene Roddenberry (1921–1991) is an American science fiction writer. He is best known for his *Star Trek* stories.

[102] George Lucas (born 1944) is an American cinematographer. He is best known for his *Star Wars* stories and movies.

as the sum total of a series of poor choices that have been made by an individual. Could there be some fundamental connection between karma and the "wrong choices" made in the context of human free will?

We Write the Stories of Our Lives in the Choices We Make

We have already discussed in chapter 4 how information and choice are intimately linked. If we think about our personal lives for a moment, it will become clear to each of us that the story of our lives is written down in the language of choice. Every day we face new choices. Some choices are easy, some choices are hard, and some represent key decisions that determine the future course of our lives. All human lives are stories written irreversibly, chronicling our choices. We can't stand apart from making these choices. Should we refuse to choose at all, it is as if we are not even alive.

Circumstances will always bring us face to face with challenges and opportunities that require us to make a choice. The business of life is making choices. As we have already seen, making choices creates information. The business of life is in fact the creation of information. We are born, we die, and in between we are constantly in the process of creating the new information. In so doing, we tell our life story. Much of our personal information will ultimately be lost because the methods we use to record information are, like ourselves, perishable.

Some parts of our story may live on in the memories of other people. Of course, we all hope to be remembered in a good way. This is why most of us try to make the best possible choices while we live our lives. However, sometimes we do not make the best possible choices, and to be honest, all of us can look back on some of our decisions with a sense of regret. When recalling these flawed decisions, we may say to ourselves, "Why did I do or say that thing? What was I thinking? How could I have been so wrong?"

Since the record of our choices is irreversible (recorded either on paper, electronically, or ultimately in another person's mind), once a decision has been made or an action has been taken, it can never be taken back or undone (the die is cast!). But poor choices are also a fundamental part of our human condition. We all make mistakes. Making mistakes is just as human as doing the right thing. At certain points along the way, our lives may get so twisted up in bad decision making (and the problems our poor decisions give birth to) that we may come to feel that it has become impossible for us to unravel our problems. We long to return to a clean slate condition where our decisions from the past—especially the ones we regret the most—have been erased. This wish is a very difficult one to have granted.

Every one of us has a past, and we all regret some part of it, in whole or in part. But what is a good decision? What is a bad decision? Sometimes good and bad choices are very clear, but in other cases, the right choice is not at all clear (the right choice may only seem clear in hindsight and maybe not even then). How do we know what is the right choice? In what direction is our moral compass pointing? How can we know the direction of true north for making a good decision?

In truth, none of us has such a compass. Often we find making a good decision is not as easy as it might sound and not in the least as easy as some people would have us believe. It is amazing how often well-meaning people have fast, pat answers to *our* problems and dilemmas. But important life decisions and major choices are always, at their root, personal choices. (As the song says, "You've got to walk that lonesome highway, you've got to walk it by yourself, ain't nobody going to walk it for you, you've got to walk it by yourself.")

We are the ones who must ultimately live with the results of these decisions and not that other person who is giving us his or her well-intentioned advice. Ultimately we all make our own choice, and we are required to live with the results. Other people may help us clarify the issues we are facing, but ultimately the choice is ours alone. How can we acquire the right attitudes to help us make the best possible choices for our lives? Let us stand back for a minute and examine what it means to make choices and reflect on the kinds of choices that are available to us.

Information and Choice

Always keep in mind that information is, by its very nature, a material thing. Even the information from God that is manifested into our world at the moment of wave function collapse is inherently and fundamentally material. Like all information, God's information is ultimately lost through

141

the process of entropy. Therefore we humans must always be involved in the process of refreshing our godly information. Otherwise the processes of confusion and degeneration begin to take hold, wiping away the memory of our personal godly encounters.

Memories of our encounters with God fade very quickly. Such memories can easily become confused in our minds. For this reason, all religions teach the importance of seeking a direct experience of God right in the here and now! This wisdom is directly tied to our human condition as material beings. We are always dealing with the reality that God's message must be, by its very nature, encoded as information and therefore is perishable. Because of the dangers of information decay, the world's religions urge us to be constantly praying and meditating, ensuring that God's message for us will always be fresh, new, and uncorrupted by the entropic process (i.e., the living water spoken of by Jesus in John 7:38).

We can encounter this very process for ourselves as we read Holy Scripture. The words of the Bible (for instance Matthew 11:28, "Come to me all that are weary and carrying heavy burdens, and I will give you rest"), the Koran,[103] or the Bhagavad-Gita[104] are all information that was recorded long ago and lovingly cherished through the generations by earnest believers.

However, the truth is that the real Scripture is what God writes anew on each of our hearts each time we hear or read the words of recorded Scripture. We all hear a uniquely personal message that is stimulated by the words on the page. But in fact, the true message is what is created for us by God in the moment! It is this kind of "in the moment" godly encounter that offers us advice we must truly rely on when making the hard choices and decisions of our life. For all of us, God lives truest and most powerfully in each and every moment of our life. A single human experience of God quickly loses its vitality with the passage of time. We must constantly open ourselves up to new and fresh experiences of God.

---

[103] The Koran is the holy scripture of Islam.

[104] The Bhagavad-Gita is a beloved holy scripture of Hinduism.

Faces of God, Laws of Science

Recently Allen and Fran visited the Asian Art Museum[105] in San Francisco, California. While there, they were treated to three floors of some of the world's greatest treasures of Asian art. On the top two floors are paintings, sculptures, and handicrafts from all over Asia. There are Chinese watercolor paintings and carved ivory. From Japan there are boxes and painted doors. There are brush-painted scrolls from Korea and sculptures from India and Southeast Asia.

Sculptures artistically depict the many Hindu gods and goddesses from India. In some cases these sculptures have as many as eight arms and hands. In other cases, the sculptures have many faces, which may completely surround their heads. To some Christians, these sculptures are simply the idols of a foreign religion. However, what we see in these sculptures is the attempt by the artist, whoever he or she was, to depict the diversity of God within that sculpture.

Such sculptures bring up for us the question of how many different ways any of us can experience God during our time on this earth. In the Hindu and the Buddhist traditions, there are numerous tales of one or more faces of God fighting on the side of humanity against the demons who threaten human freedom. It should be pointed out that the depictions of God in the Asian sculptures that were encountered by us in the Asian Art Museum are very similar to forms of God depicted in the Bible:

> Then I saw between the throne and the four living creatures
> and among the elders a Lamb standing as if it had been
> slaughtered, having seven horns and seven eyes, which are
> the seven spirits of God sent out into all the earth. He went
> and took the scroll, and the four living creatures and the
> twenty-four elders fell before the lamb, each holding a harp
> and golden bowls full of incense, which are the prayers of
> the saints (Rev. 4:6-8).

In these Asian cultures, there is a tradition of God (in God's many forms and many names) fighting for humans against the demons who

---

[105] The Asian Art Museum of San Francisco has one of the most complete collections of Asian art outside of Asia.

are trying to bend people to their will by robbing them of their freedom. Perhaps we could associate these demons with our perennial enslavement to those raw survival emotions from our own evolutionary history. As with any addiction, an individual's freedom is lost when the addiction takes hold. But how does God rescue people from these demons?

The answer is by the transformation of emotions like anger and fear into God's own virtues of compassion, hope, and trust. When we are despairing and asking God to send us help—and help in the form of hope—then we are asking for such a transformation. When we feel guilty because we have done something wrong, then we pray to God to send us mercy and forgiveness. Like hope, mercy and forgiveness are also a transformation. If we are in a position to help someone but we think we are too busy with our own affairs to be able to help, and God sends us compassion and courage to change our priorities; this is transformation.

In the Chinese tradition, the face of God that in the Christian world is called mercy is known as *Kwan Yin*. In Japan, the face of God that is called compassion is named *Kannon*. By whatever name we call them, God's virtues, whose essence is beyond words, is only understood as creatures of the heart. God's virtue is the answer to our prayer for freedom from the compulsion of our raw survival emotions (such as anger, greed, and fear).

Another face of God that grants us insight is called intuition. What we learn from these experiences of God is that the loving kindness of God's grace is bestowed upon us in the form of God's many virtues. God is hope, God is mercy, God is compassion, God is forgiveness, God is trust, God is beauty, God is wisdom, and especially God is *truth*.

Laws of Science and Laws of God

Could it be that the laws of science are also among the faces of God? One of the characteristics that distinguishes human beings from other animals is our unique ability to understand, control, and master our environment through what we have come to call science and technology. From the first crude stone knives to today's supercomputers, humans have always possessed the ability to fashion tools that are needed to adapt to and master new and changing environments. This ability has enabled human beings to spread to all corners of the earth.

Other animals must wait for many generations before the natural selection process brings about the physical changes to their bodies that allow them to adapt to environmental change. But humans, by having the ability to understand and control their environment (by way of their knowledge of science and technology) very quickly become its master. As physicist Stephen Hawking[106] put it: "The earth is an unremarkable planet orbiting an unremarkable star in a little corner of an unremarkable galaxy, but we humans have the ability to imagine and understand the workings of the entire universe."[107]

Let us now consider the question of whether the laws of science are simply a changeable part of our material universe or if they are in reality facets of eternal truth. If they are indeed facets of the one truth, then the laws of science must also be faces of God. Therefore, those who seek to understand the laws of science (i.e., scientists) are also seekers of God whether they recognize it or not! The meaning of the word *God* and the word *truth* are the same (John 18:37).

At the moment of the big bang's singularity, all matter and energy that is and ever will be within our universe came into being. Many people assume the laws of physics and chemistry (and much later when life first appeared on earth, the laws of biology) came into being at the same moment as the big bang, but can that really be true? Did the laws of science have a beginning? This is a very hard question to answer because without the presence of matter, energy, space, and time, the laws of science would have nothing to work with and would represent an unobserved potential to action rather than action itself. The laws might be mute, but they could still exist! This question really asks that we have an understanding of what was going on inside of God *before the big bang*! Questions like this are way beyond human understanding. But it is a valid question, humbly asked.

Let us try another approach. Are the laws of science unto themselves creatures of matter, energy, space, and time? The answer is a resounding *no*! They are principles by which all matter and energy have predictable behaviors during their interactions as they move through space and time. We think of the laws of science as being expressed in mathematical

---

[106] Stephen Hawking (born 1942) is a British physicist and cosmologist best known for his theoretical work on black holes.

[107] Stephen Hawking, *Black Holes and Baby Universes and Other Essays* (New York: Bantam Books, 1993), 50–53.

formulas. But the principles themselves are not mathematical formulas and are not even numbers. These mathematical expressions of the principles are simply ways in which humans express the laws of science using the only language that is understandable to us. The mathematical expressions of the laws of science are completely analogous to how humans are able to express their emotions by using the languages of art, poetry, music, and dance. If the virtues are the faces of God embodied by human beings undergoing emotional transformation; then why are not the principles of science, by analogy, the faces of God, likewise embodied by human beings in their quest to understand the natural world we know as our universe?

Some people will say, "The laws of science change as science progresses. How can something as changeable as the laws of science possibly be associated with God, who is eternal and unchanging?" This is a good and fair question. The answer is in the recognition that the laws of science are like babushka dolls,[108] one enclosed by another and so forth.

For example, the classical mechanics of Newton is completely contained in the general relativity of Einstein. For everyday distances and nonrelativistic speeds—in our everyday world of cars, airplanes, space shuttles, and moon rockets—Newton's mechanics works just fine. However, things are different when we analyze the behavior of whole galaxies spinning outward from the big bang at velocities approaching the speed of light. As they spread out over distances of billions of light years, Newton's mechanics are forced to defer to the more general laws of general relativity.

This situation in no way *invalidates* Newton's mechanics within their domain of application. We are simply forced to recognize that Newton's mechanics are a *special case* of a more general principle—an approximation that lacks sufficient accuracy over intergalactic distances and relativistic velocities. The Newtonian babushka is entirely enclosed within the Einsteinian babushka. The babushka doll understanding of scientific law is discussed in far more detail by physicist Steven Hawking in his book *The Grand Design.*

At the other (i.e., the small) extreme on the distance scale, a similar situation can be said to apply as Newtonian mechanics is generalized to quantum mechanics on the one hand and Maxwell's classical

---

[108] Babushka dolls were originally called matryoshka dolls, which means "little mother" in Russian.

electrodynamics is generalized to the quantum electrodynamics of Richard Feynman on the other. It has always been the mission of science to proceed from the particular to the general. However, once the general law has been identified, it in no way invalidates any of the particular expressions of the law within their limited *domain* of applicability. This constant movement from the particular toward the general is in truth the ultimate quest of science and scientists.

On the morning Albert Einstein was found dead at his home in Princeton, New Jersey, a paper containing scrawled equations was found next to him on his night stand. His final thoughts were occupied with the details of what Einstein called the universal field theory. Einstein had hoped his universal field theory would encompass the babushka of general relativity and the babushka of quantum mechanics. To this day, the quest for a fundamental principle containing both general relativity and quantum mechanics remains the unidentified holy grail of physics. The scientist's quest for the most general scientific principles exactly parallel's the theologian's (and the mystic's) quest for God. We believe these two seemingly separate quests are in truth one and the same quest. Truth is truth no matter what its name or where we find it. At least some of the faces of the multifaced statues need to be associated with the laws of science.

The Synchronicity of God's Actions: *Kairos* vs. *Chronos*

We learned in earlier chapters (chapters 3 and 4) that God acts within our universe by causing God's divine information to come into material being. God's will is expressed as a manifestation—as a choice made by God—at the moment of quantum wave function collapse. As an expression of God's omnipotence, each divine choice might be coordinated (as God acts across both space and time) with thousands, millions, billions, or perhaps trillions (or perhaps an uncountable number) of other divine choices, thus providing a richly detailed macroscopic mosaic expression of the divine will.

We call this coordination of divine choices quantum synchronicity, and as mentioned in chapter 4, this harmony of choices is very similar in its function to what C. G. Jung and Wolfgang Pauli called synchronicity.

But how might these harmonized godly choices be manifested within our conscious human mind?

Consider the human central nervous system that is made up of billions or perhaps trillions of neurons and synapses. (One estimate is that the total number of synaptic interconnections within the human brain is approximately equal to the total number of galaxies in the universe, over 100 billion!) The synapses in our brain are continuously firing and processing bits of electrical and chemical information, which adds up to total brain and nerve function. Perhaps the harmony of God's creative choices is the process by which the positive emotions of love, compassion, kindness, gentleness, hope, joy, and truthfulness are implanted to our consciousness.

On the other hand, our negative emotions such as hate, fear, greed, and lust are all born out of our genetic past—a legacy handed down to us from our animal ancestors. These negative emotions are absolutely necessary for our protection and survival, but they are easily abused in the everyday circumstances of living as social creatures. Many, if not most, of the life choices we face as human beings come down to discerning what is a positive emotional response and what is a negative emotional response in a given situation. Often by choosing the negative emotional response we follow our own will, whereas by choosing the positive emotional choices, we follow God's will for us. The Bible speaks to us about emotional choices in what are thought to be the words of St. Paul: All of us once lived among them in passions of the flesh, following the desires of flesh and senses, and we were by nature children of wrath, like everyone else. For by grace you have been saved through faith, and this is not your own doing! It is the gift of God, not the result of works, so that no one may boast. For we are what he has made us, created in Christ Jesus for good works, which God prepared before hand to be our way of life (Eph. 2:1-10).

To the ancient Greek philosophers, the moment in which God's will is received by us is called *kairos* (as opposed to *chronos*, which is the Greek name for the everyday timekeeping by clocks). We can't escape making difficult emotional choices. People who make a habit of choosing negative emotions can be said to be walking on life's dark side. We must always remember that when we feel a negative emotion welling up inside of us that we are in no way compelled to embrace this memory from our distant genetic past.

The Shark

Allen writes: In my home state of Rhode Island, a large and very beautiful bay extends inland from the Atlantic Ocean for almost the full length of the state. In the middle of Narragansett Bay is a pair of small islands called Patience and Prudence islands, which are named for the daughters of the first English settlers to the area in the seventeenth century. I can remember one night, at dusk, being out on the bay fishing with my dad. We were anchored between Prudence and Patience islands and fishing for flounder.

All of a sudden I felt a terrific pull on my line and my fishing pole was nearly ripped out of my hands. I was sure that I had hooked a whale! I fought the fish for about fifteen minutes and finally got it to the edge of the boat. With one final heave, I pulled my "fish" into the boat. But instead of being the size of a large flounder, which I expected, it turned out to be a very angry, very large shark.

The shark flopped around in the bottom of the boat, exposing his very large, very sharp teeth. I just wanted the shark out of the boat. Using a pair of pliers that I was fortunate to have in my tackle box, I was able to remove the hook from the shark's mouth as he tried his best to bite off my fingers. I grabbed the shark by the tail (far from the shark's teeth) and, using every ounce of my strength, flung him back into the water. I was shaking all over with fear of what might have happened if he had gotten hold of my hand. I am not sure if patience is an emotion, but I am very sure that fear (of sharks, at least) is.

As this story demonstrates, inherited survival emotions from our genetic past can be lifesaving (in this case, as much for the shark as for Allen). These emotions give rise to the "fight-or-flight" response, which can prove lifesaving under dire circumstances. These survival emotions only become problems for humans when they are misused and abused. The following stories make this point in the telling of a number of individuals' experiences.

The Parable of the Rug Master

It is a part of our humanity that we all make poor emotional choices from time to time. However, because of God's infinite wisdom, God often finds a way of transforming our mistakes in the long run into something good and beautiful. The following is a parable from long-ago Persia describing the wonderful creativity of God's transformative actions.

Long ago in a country that is now called Iran, when Persian rugs were still made from the natural dyes derived from local plants, the process of rug making proceeded in the following way. As it was taking shape, the rug would be hung from a vertical frame. In back of the frame were several tiers of benches, one bench tier above another. Small boys sat on these benches. Each boy had in front of him several balls of thread, and each thread was a different color. By attaching a particular thread to a long needle, the boy could push the thread of his choosing through the rug to loop it around to the front side of the rug and then come back around again so it could be tied off.

None of the boys could see the rug until it was finished. Only the rug master could see how the rug was growing and maturing at each step in the process. When the rug master asked a particular boy to pass a thread of some color through the rug, the boy would pick up his needle, put the thread of the desired color into the needle, and pass the needle and thread through the rug at the place where the rug master had indicated. Only the rug master could see the entire rug, so the boys had no idea how their work was progressing.

However, like small boys everywhere, sometimes a boy would become distracted by something his neighbor did or said. When this happened, the boy might make a mistake and not do what the rug master asked him to do. It is here that the artistic genius of the rug master really became apparent. For rather than letting the boy's mistake(s) detract from the overall value of the rug, the most-talented rug masters would incorporate the boy's error into the overall pattern in such a way that the error itself would become an unexpected feature of the rug's overall plan. It was these very rugs, with their unusual and unexpected patterns, that would become the most valuable and most highly sought after rugs, and they would bring the highest prices.

Our lives are like Persian rugs. Our raw survival emotions are like the small boys, and God is our rug master. Sometimes we make mistakes because our emotions may lead us astray. But God knows our natures, with all their blind spots and shortcomings, and when we make a mistake, God, if we will only let him, will find a way to help us weave our mistakes into the overall pattern of our lives. Like a Persian rug, the beauty and wisdom of our lives will ultimately be enhanced and transformed. If we let him, God will transform our negative emotions into virtues. With God's help, our fears become the seeds of our courage, our selfishness the root of our generosity, and our anger the beginning of learning patience. It is by God's creative wisdom that our hardwired emotions of survival will be used like the small boys of the rug master story to weave unexpected patterns of great beauty into the stories of our lives.

Next we will encounter a story from Japan illustrating again how God's wisdom can find a way to transform raw human emotions into something of great and lasting beauty.

The Story of a Zen Buddhist Novice

This is the story of a young Zen Buddhist novice from a Japan of long ago. This story is very similar to the gospel story (Matt. 19:16-26) of the rich young man who was not yet ready for discipleship. This young man was a serious student of the Buddhist scriptures, and he was quite proud of his scholarly achievements. He went from monastery to monastery looking for a master who would appreciate his talents.

He finally went in search of a particular master who had been recommended to him by a number of older monks. When he found the master, the young novice lost no time in trying to impress the master with his amazing grasp of Buddhist scriptures. The master's only comment was that the young man's knowledge was quite interesting, but he would accept him as his novice anyway.

However, there was trouble between the two of them from the start. At every meal, as the novice arrived at the communal table, the master would say very loudly so all could hear, "Here is the monk who knows everything there is to know about holy scripture." When the two would pass on walks in the courtyard, the master would say very loudly, "This monk knows everything there is to know about the fine points of holy scripture." The other monks would hear what the master was saying, and sometimes they would laugh. Pretty soon the novice became what he thought was the laughingstock of the monastery.

One evening he could stand it no longer. The young novice resolved to confront his master over the outrageous things he had been saying about the novice. The young novice went to where the master lived and found him sitting on his porch, sipping a cup of tea. He walked up to his master and said, "I want you to stop saying the untrue things you have been saying about me in public." The master laughed and said, "Is it not true that you are a fine scholar of holy scripture? Isn't this what you have been telling me all along?"

The young novice was caught off guard by the truth of this remark and threatened to strike his master. But the master struck first, knocking the young man off the porch. Next the master jumped down from the porch, and the two rolled over and over, fighting in the dirt and the mud of the courtyard. The entire monastery heard the commotion and came out to see what was going on. As the fight continued, the whole community looked on.

Just as soon as the fight started, it ended because the young man woke up to what he was doing. He got up and said nothing but walked silently to his room and packed his few possessions. Without even a word of good-bye, the novice left the monastery forever. How could a novice who fought with his master ever hope to be a part of this monastic community or any other community, for that matter?

The young man wandered aimlessly along a river many miles from the monastery. Finally, to escape the cold and wet winter weather, he made a crude place for himself in a cave by the river. He sat in his cave for days at a time in a shocked stupor about what he had done. When he became hungry, he walked into a nearby village where it was a tradition to feed the hermit monks who lived nearby. The villagers put enough rice in his begging bowl so he didn't starve. But he spent most of his time back in his cave, just thinking about what he had done.

In time his mind turned to the scriptures he had been so proud to be a scholar of. But now he began to ask himself what the Buddha's words really meant. How did the Buddha's words affect him personally? What was it he needed to hear? In time he began to get answers. Slowly at first, he began to feel the intense meaning of Buddha's words for himself. The Buddha's words were no longer simply an intellectual exercise, as they had been for him in the past. He now saw clearly that the Buddha's words were meant for him.

In time, which was measured in decades, the monk truly achieved his own enlightenment. By the time he was an old man, word had spread across this part of Japan that a hermit monk who lived in a cave by a river had truly found enlightenment. Perhaps a very lucky novice or two might learn the way from this wise old Zen master.

The Story of the Man Who Nurtured Hate

This is the story of a man who had the misfortune to be a small child growing up in Greece during World War II. A battle occurred near his small town, and the victors of the battle herded all of the townsfolk into a square and began questioning them to ferret out spies. For some reason, these solders thought the boy's mother was a spy and shot her on the spot, right in front of the boy's watching eyes. Somehow the boy managed to escape with his life. Later, a family member helped him come to America and live with relatives in Chicago.

But the boy never forgot the horrible brutality of his mother's death. He swore that someday he would return to Greece and have his revenge by killing his mother's murderer. When the boy grew up, he worked hard and hired a private detective to go to Greece and find the man who killed his mother. The detective really did find the boy's mother's killer. The man who killed the boy's mother was now retired and living in a small mountain village. The detective gave the man in America the address and detailed information about the old man's comings and goings.

Here at last was his chance for revenge. The man in America had dreamed hateful dreams of revenge for most of his life. All of the man's hatred for his mother's killer, which he had carefully nurtured over so many years, was now focused on his ultimate mission of killing his mother's killer. He purchased a weapon and made airline reservations, and in a matter of days, he was back in Greece, the land of his birth.

Since the town where the old man lived was in a remote and mountainous area, it took the man from America several days to reach the little town, but reach it he did. The man from America checked into the only hotel in the town, unpacked his weapon, and checked his ammunition for the final mission that was to come.

He wasted no time in finding his quarry. But when he reached the killer's house, what he found was a very old, very infirm man sitting alone in his backyard among his grapevines and roses. This picture did not match at all with the man's childhood memories of brutality and murder. This old man looked nothing like the horrible killer who murdered his mother so many years ago. What he looked like most was an old man nearing the end of his days.

Nevertheless, it was him; the detective had provided proof of his identity beyond any reasonable doubt. So the man from America took out his weapon and prepared to do the killing he had been planning and dreaming of for most of his life. But one thing was missing: his *hate*! He couldn't muster that boiling, white-hot hate he had nursed for all those years in America. Without the rising of his hatred, he couldn't do the killing. Instead, he realized that in fact he would be doing a terrible thing by killing this old and sick man who seemed so close to his own natural death.

The man put away his gun and returned to the hotel, hoping that in the privacy of his hotel room he might be able to conjure up the burning, white hot hatred that would be required to complete the act. But he couldn't, though he tried all night and all the next day to conjure up his hot, boiling hatred. Without his own personal hatred, he began to realize that all he would be accomplishing would be the murder of an old and sick man—a crime for which he could expect to spend the rest of his own life in jail. He waited one more day, but his hatred never returned, so he reluctantly packed his bags and returned to America, grateful that something or someone (i.e., God's grace) had saved him from crossing over that line and becoming, like the man of his horrific childhood memory, a murderer!

Emotional Addiction

The emotional legacy of our animal ancestors protects us from many of the dangers life dishes out. Our fears register caution in threatening situations, our anger helps us to face attack, and our lust helps us to find and attract a mate. However, in extreme situations these hardwired emotional responses, based on survival instincts alone (inherited from our evolutionary past), can become so overpowering that they will literally take possession of our lives. If an individual becomes so habituated to a particular emotion that the emotion begins to control his or her life, it may be that this individual has become addicted to the emotion, just as one might become addicted to a substance like alcohol or cocaine. Like substance addiction that leads to a downward spiral of out-of-control behavior, emotional addiction too will often lead to its own kind of out-of-control, downward spiraling behavior.

A tragic example (with a hopeful ending) of emotional addiction is the story of a platoon of American soldiers serving in Iraq during the surge campaign of 2007–2008. This platoon suffered massive losses during a particularly horrible patrol, causing most of its surviving members to fall into an emotional tailspin of grief, anger, and lust for revenge.

En masse, the troubled platoon bravely made their way to their unit's mental health facility located near their quarters. The doctors talked to the men of the platoon and suggested they stand down from patrol for a while because in their present state of mind, they might find it impossible to control themselves during threatening situations. The men themselves admitted to having a fear that if they went out on patrol, they would kill anyone who crossed their path, including innocent civilians and perhaps children.

Out of the fear that they might do something they would regret for the rest of their lives, the platoon, to the last man, refused to obey an order from their commander to go out on patrol. The men of the platoon were accused of mutiny by their commanding officers. Each man was stripped of all his privileges, and all were sent home to the United States within three months. Although they were accused of disobeying orders, their acts of courage in standing down from patrol while deep in the grips of an emotional addiction perhaps saved many innocent lives and the future mental health of the soldiers themselves.

CHAPTER 8

THE POSSIBILITY OF
EXTRATERRESTRIAL LIFE

Fermi's Paradox

Italian physicist Enrico Fermi[109] was working at the Los Alamos National Laboratories[110] in New Mexico in the spring of 1950. One day, Fermi and three of his colleagues were enjoying a long, leisurely lunch at the laboratory's dining room. During lunch, their conversation turned to a discussion of the possibility of extraterrestrial life. All four men enthusiastically agreed that surely our universe must be teeming with life. After all, it made no sense at all to conceive of our earth as the only island of life that exists in an otherwise sterile universe. Fermi and his colleagues all agreed that the probability against such an unlikely scenario was surely a million or even a billion to one.

After lunch, as the four physicists were walking back to the work site, the quiet, introspective Fermi remarked to his colleagues, "If there is so much life out there in the universe, where is the evidence of its existence?"

---

[109] Enrico Fermi (1901–1954) was an Italian physicist best known for his work on solid-state physics. Together with Leo Szilard, Fermi built the world's first self-sustaining fission nuclear reactor at the University of Chicago at the start of WWII.

[110] Los Alamos National Laboratories was constructed early in WWII to house the atomic bomb project in a very rural part of northern New Mexico. Today the laboratory remains a key location for conducting high-energy physic research for national defense.

The other three just shrugged their shoulders because no one had a good answer to Fermi's question.

Fermi's question, which remains unanswered to this day, has come to be known in cosmology circles as Fermi's paradox. Even today there are no readily available answers to his question, nor is it even clear how one might go about obtaining an answer. An answer may have to wait for the occurrence of some unpredictable event. For the people of our time, the confirmation of an extraterrestrial contact might carry the same impact as what was experienced by Jesus's disciples during their mountaintop experience of the transfiguration.

The Evidence of Extraterrestrial Life

What kind of evidence would convince us of the reality of extraterrestrial life? Of course face-to-face contact (assuming extraterrestrial beings have faces!) would be best, but this is unlikely to occur given the great distances that are involved with interstellar space travel. Furthermore, it is difficult for any of us to even conceive of a technology that would enable travel across such vast distances (hundreds, perhaps thousands of light years). Perhaps a plan to use some form of signaling to establish contact with some technology-savvy extraterrestrials would be a more plausible approach.

Communications engineers understand that the best solution to the problem of signaling across the vast distances of interstellar space is to use electromagnetic waves in the part of the radio spectrum that is called microwaves. (This is in approximately the same frequency range as the cosmic microwave background noise left over from the big bang and where earthbound cell phones operate.) If we only could receive a microwave signal from planet X, which is orbiting star Y in galaxy Z, we would have an answer to Fermi's question, and humanity would finally know once and for all that we are not alone in the universe!

But wait one minute; this intergalactic eavesdropping may not be as simple as it sounds. You can't just tune your favorite radio to receive microwave frequencies while pointing your satellite dish antenna at some star in the night sky. Let us step back for a moment and see what it might take to receive these distant signals based on the laws of physics when they are applied to communications engineering.

To send and receive greetings over the vast distances of interstellar space, the following steps must be taken. First we must assume that someone, somewhere (probably not us here on Earth given the status of our present-day technology and our economic situation) must be eager to send out a cosmic greeting with sufficient power to span these vast interstellar distances.

The first problem these people (assuming they resemble us) must face is what direction to point their antenna to be certain their signals will reach their intended listeners. This is a serious problem because in all likelihood they have no idea where their listeners are or they wouldn't be forced to send out beacon signals advertising their presence. The only reasonable solution to this problem is for the radio engineers of this civilization to transmit their signals equally in all directions, using what is called an omnidirectional antenna. However, omnidirectional antennas (as opposed to truly directional antennas like parabolic dishes) are very wasteful of transmitter power because they radiate equally in all directions, spreading their radio energy far and wide so that only a small percentage of the total transmitted energy ever reaches the unknown target's location. However, let us assume that planet X's microwave radio signals are of sufficient power to ultimately reach us here on Earth. What kind of transmitted signal power is necessary for us here on Earth to receive planet X's signals?

All electromagnetic waves (including radio waves) decrease in intensity as the distance from the transmitter is increased according to what is called (in classical electrodynamics) the inverse square law. The inverse square law tells us that if the range from a transmitter is doubled, the received signal strength will decrease by a factor of four (i.e., two squared). Even over distances within our solar system (as in the case of an Earth station receiving signals transmitted by space probes sent out from the Earth) signals transmitted by probes from well beyond Earth's orbit are of nearly imperceptible signal strength when they are received back on Earth. Very large directional antennas (such as arrays of large parabolic dish antennas) are needed to collect sufficient signal power to enable the extraction of useful information from these signals.

Communications engineers define a parameter called the signal-to-noise ratio of very weak signals as the measure of how detectable the information carried by these weak signals really is. Often sophisticated receiver systems make use of very narrow—bandwidth radio frequency filters to increase the signal-to-noise ratio of extremely weak received signals (since the noise power of a receiver is proportional to the receiver's bandwidth, decreasing the receiver's bandwidth increases the signal to noise ratio) at the expense of the signal's data rate. This was predicted by Claude Shannon's famous equation for the information capacity of a noisy channel (see chapter 4 for details).

By exploiting this natural tradeoff, NASA scientists have been able to transmit, and receive on Earth, wonderfully detailed pictures of the planets Uranus and Neptune from the very edge of our solar system. To fully exploit this technique, the data rate of the information being transmitted from Neptune had to be slowed way down to the point that it took almost twenty-four hours to transmit each one of these spectacular pictures! However, improving the signal-to-noise ratio of a very faint signal by reducing the receiver bandwidth must always be purchased at the expense of data rate. Unless all we are interested in receiving from planet X is a brief, "Hi, we are out here," the reception of very low signal-to-noise ratio signals from deep space is of limited usefulness for communicating something really significant about a distant civilization.

Another way to look at the transmission of microwave radio signals over extreme distances involves using the concept of the cosmic microwave background radiation left over from the big bang's fireball as the limiting factor that determines the ultimate signal-to-noise ratio of an extremely weak signal from deep space. This cosmic background radiation is encountered no matter what direction the antenna is pointed in the sky. Communications engineers describe this cosmic background radiation as having a noise-equivalent temperature of 2.7 degrees Kelvin. Noise temperature is the concept used by communications engineers to specify an equivalent overall temperature that is associated with the noise created by electronic devices, such resistors or wires (that comprise the receiver's circuits) when these elements are heated to some ambient temperature that is above absolute zero.

As we learned in chapter 4, every bit of information must contain at least a minimum amount of energy that is equal to Boltzmann's constant (1.38×10^{-23} joules/degrees Kelvin) times the environment's ambient temperature in degrees Kelvin. A faint signal beamed toward Earth must have, as a bare minimum, an energy per bit of at least Boltzmann's constant times 2.7 degrees Kelvin for the signal to be distinguishable above the constant hiss of the microwave cosmic background noise.

From a practical viewpoint, this is truly a bare minimum requirement since any additional noise sources associated with the receiving equipment itself, or with the environment where the receiver is located, will raise the signal's minimum required energy per bit. All that can be done when receiving extremely weak and barely perceptible signals is to increase the time duration of each bit. This will increase the energy per bit and therefore increase signal-to-noise ratio. Such a process is mathematically equivalent to the technique for reducing the receiver's filter bandwidth, as mentioned above. Yet again we see that signal-to-noise ratio must always be purchased at the expense of data rate. This principle is an iron law of communications engineering.

If planet X is located at some distance away from Earth, the received signal power on Earth must be maintained above some minimum level if the signal's data is to be successfully extracted at a useful rate. Remember, the distance from the transmitter to the receiver always translates into the signal strength declining according to an inverse square law. Given all of these natural conditions and constraints on the reception of extraterrestrial signals, the only factor for increasing reception ability that can be controlled on Earth is increasing the antenna's gain. (Gain is a technical term describing an antenna's directional ability.)

Let's now go to an extreme and estimate (based on the factors discussed above) just how deeply into space we could ever hope to receive a radio signal. First of all, we must estimate the maximum feasible transmitter power generated on planet X. Going for broke, we assume the people on planet X who are building and operating this transmitter are willing to devote sufficient resources to this project for their transmitter to be capable of producing a signal power output that is roughly equivalent to 1/10 of the total solar power output (i.e., our sun's energy) that is constantly falling on the face of the earth (the solar flux at the earth's surface is about one kilowatt per square meter) during daylight hours. This is a truly astronomical amount of transmitted signal power and represents many orders of magnitude more signal power than we could imagine generating here on Earth.

The second concern is how much antenna gain could conceivably be associated with an Earth-based receiving station? Let us assume this antenna can be pointed at any point in the sky. Since we don't know where planet X is, we are forced to search for its signals by sweeping the antenna progressively across many different parts of the sky. However, the best place to locate such a large antenna is not here on Earth but on the far side of our moon. The mass of the moon would act as a natural rock shield protecting the receiving antenna from the cacophony of radio interference generated by all the heavy machinery and cell phones operating on Earth.

Assume for the moment that we have at our disposal an astronomically large antenna located on the far side of the moon. Since the antenna's size determines its gain, let us go for broke and assume this antenna actually occupies 1/10 of the total surface area of the far side of the moon. Given these two conditions and the additional assumption that the data rate of the received signal must be maintained at a rate at least as fast as a normal broadband Internet data line connection (i.e., about 1.5 MBs), we can show with a very straightforward calculation based again on Shannon's law for the information capacity of a noisy channel that the maximum range over which we can ever hope to detect a radio signal is between one hundred and two hundred light years.

This is a very interesting result for a number of reasons. First, because radio signals travel at the speed of light, we are talking about a one-way trip for the radio signals of about one or two human lifetimes. Second, to maintain a sense of proportion, we must recognize that one hundred to two hundred light years is just a fly speck in the vastness of intergalactic space. (Consider that our entire universe is well over 27 *billion* light years in diameter). One hundred to two hundred light years is a very modest distance that is well within the local neighborhood of our own little corner of the Milky Way galaxy. This calculation really tells us that unless our local corner of the galaxy is teeming with life—that is so densely packed to have two (we on Earth are one of them) highly technological civilizations capable of long-distance radio signaling and located within one hundred to two hundred light years of each other—we may be permanently and fundamentally out of communications with even our closest technological neighbor. The universe may indeed teem with life, but we here on Earth may never be able to experience it!

The Drake Equation

Another extremely interesting way to look at this issue is to consider the Drake equation.[111] This equation, named for cosmologist Frank Drake, provides a way (however speculative) of estimating the density of intelligent, technologically advanced civilizations within our universe. Of course there are a lot of the specialized numbers that must be plugged into the Drake equation, all of which are really pure guesswork. For instance, how do you estimate the number of technologically advanced civilizations with an ability to create nuclear weapons and destroy themselves that exist within some given period of time?

However, even given all of its shortcomings, the Drake equation is still the best and only way of estimating what could be a reasonable distance between Earth and the nearest star/planet that is capable of transmitting radio signals we are capable of receiving. Using the best guesses available at present for these factors, the Drake equation predicts the spacing between technologically advanced civilizations within our Milky Way galaxy to be about two thousand light years! (This is based on the prediction of a total of two thousand technological advanced civilizations within our Milky Way galaxy.)

This spacing between technological civilizations is a factor of ten greater than what we have calculated to be the maximum radio signaling range about planet earth. Again the result strongly suggests that radio communications with another civilization, or perhaps other civilizations, within our galaxy is *not* possible today, and may never be possible, based on fundamental laws of physics.

---

[111]   The Drake Equation was first developed by cosmologist Frank Drake to provide a very crude estimate of the density of planets within the universe that contain technologically advanced civilizations. Many of the factors in the Drake equation are admittedly crude guesses. However, at this point in our understanding of extraterrestrial life, the Drake equation is the best estimate we have.

SETI

For many years, the SETI (search for extraterrestrial intelligence) Institute[112] in Mountain View, California, has been listening for radio signals transmitted by extraterrestrial beings in interstellar space. SETI uses highly directional antenna arrays and ultrasophisticated radio receivers to survey the sky both in terms of a sky direction and radio frequency. The SETI surveys have been exhausting both in terms of the length of time they have been in operation (decades) and the full extent of their direction and frequency surveys. So far the results have been completely negative.

One of the biggest problems confronting the SETI team is terrestrially generated radio interference. The interference problem is becoming more and more difficult to deal with as the worldwide popularity of cellular phones, Wi-Fi computer-to-computer data communications, and Bluetooth radio links (and their technological descendants) increases. Unfortunately, these electronic communications devices operate at almost exactly the same frequencies that are ideal for interstellar signaling. Although the SETI team remains undaunted and continues to develop progressively more sophisticated radio receivers, their quest may have a low probability of success considering that nearly 70 percent of the world's population now carries cell phones!

What Could Our Isolation Mean?

Perhaps the situation that we are encountering in our search for extraterrestrial intelligence is in keeping with God's plan. We know from our experience with human history that when a more technologically advanced culture comes into contact with a less technologically advanced culture, the more-advanced culture always takes over the less-advanced culture in more ways than one. Consider what happened to the Native American cultures when the first Europeans arrived in the western hemisphere. Perhaps our Earth-bound culture is being insulated from

---

[112] The SETI Institute is located in Mountain View, California, and was founded by cosmologist Frank Drake. It is dedicated to the search for extraterrestrial life. Most of the work that has been done so far consists of sky surveys of microwave radio signals from outer space.

other more-advanced cultures of the greater universe to ensure that our culture grows independently into what it is supposed to become according to God's plan. Perhaps the reason for such great astronomical distances within our universe is simply God's way of spacing out the developing life of evolving cultures, which is a way that ensures their continuing unimpeded development.

The EPR Paradox, Yet Another Possibility for Contact across Vast Cosmological Distances

In 1935 Albert Einstein proposed a Gedanken experiment that was his final attempt at casting doubt on the validity of the Copenhagen interpretation of quantum mechanics. Together with two of his young colleagues (Boris Podolsky and Nathan Rosen) at Institute of Advanced Studies in Princeton, New Jersey, Einstein envisioned a thought experiment in which two particles were initially prepared in a single quantum state. Sometime later, the two particles would fly apart, becoming separated by a considerable distance. If a scientist were to measure the first particle, he or she would also be measuring the second particle because the two particles were initially prepared as a single quantum state and the properties of each were intimately linked to the other so the second particle's properties would become knowable (i.e., measured) when the first particle was measured! If the first particle was measured, the second particle would in effect be simultaneously measured, no matter how far apart the two particles were.

Here is the paradox: when the first particle is measured, in effect the second particle has also been measured *at the same moment*, no matter how distant one particle is from the other particle. The separation between particles could be billions of light years; it doesn't matter! But Einstein's theory of special relativity forbids information from being transferred in times shorter than the distance between the particles divided by the speed of light. What if the two particles were separated by the entire width of the universe? Einstein's theory of relativity would say that it should take billions of years to transmit the measurement information from one particle to the other. But the EPR experiment suggests this information would become available instantaneously!

To Einstein, the EPR experiment was pure quantum nonsense, casting grave doubts on the Copenhagen interpretation of quantum mechanics. However, quantum mechanics is not so easily cast aside. The EPR effect, which is now called entanglement by most scientists, is very real and in fact forms the basis of the emerging science of quantum computing. Instantaneous signaling over vast distances may well be one of the most difficult aspects of quantum mechanics for us to comprehend, but it is real nonetheless.

Consider how entanglement might facilitate interstellar communications. At the time of the big bang, if the laws of physics applied at all, then it is very likely the infinitesimal singularity that would grow into our universe would reside in a single quantum state. This condition would correspond to a total entropy of the infant universe being exactly zero, because the logarithm of a single quantum state is exactly equal to zero (i.e. $\log (1) = 0$). This is another way of saying that our universe was born in a state of perfect order. Disorder (as measured by growing entropy after the big bang) did not begin until the single quantum state of the singularity began to divide into multiple quantum states as the infant universe expanded. But all of these new quantum states were entangled with one another because they all had been prepared simultaneously within the genesis quantum state of the big bang's singularity. The Lord our God is one God (Deut. 6:4)!

Fast forward to the present day. Every quantum state within our bodies is in some way entangled with other quantum states scattered throughout the universe. If we were to perform a measurement of the quantum states within our own body (perhaps through some form of meditative exercise comprehending the state of our brain's individual synapses), it might be possible for us to train our mind to comprehend the entangled information of our twin quantum states that are spread across interstellar space. Of course it is highly speculative to even guess where this process might lead, but we can say with some degree of assurance that the laws of physics have at least left this small crack in the door open for us to ponder the possibility of instantaneous communications across vast distances using only our thoughts. It is truly mind boggling to contemplate that perhaps an extraterrestrial being a billion light years from earth might be seeking ways to establish contact with us.

Within our experience and traditions, there are numerous examples of the noncausal transferring of thoughts (perhaps by quantum entanglement).

Consider how individuals of some migrating bird species find their way across thousands of miles of unfamiliar territory while making their first migratory trip. Also consider the gospel story of Thomas and Jesus (John 20,24). Thomas was not present when the resurrected Jesus presented himself to the other disciples. Days later, Jesus appeared again when Thomas was present to provide Thomas with the proof of his resurrection and invisible presence. Because Thomas had so desired to be present several days earlier, in so doing, Jesus succeeded in breaking the bounds of time and space to provide Thomas with the proof he needed.

CHAPTER 9

MYSTERY AND PARADOX

You might think of a mystery as being a beautiful well in a desert oasis in much the same way Antoine de Saint Exupery described in his book *The Little Prince* ("What makes the desert beautiful is that somewhere it hides a well . . ."[113]). Religious people dip their buckets into the well's life-giving waters for nourishment. On the other hand, you might think of a paradox as being a confusing signpost that seems to direct the traveler in two directions at once. A scientist confronting a paradox is baffled by the contradictions and struggles to make sense out of the confusion. In a moment of brilliance, a scientist may recognize another way (not acknowledged by the sign post) to the destination that is short, simple, and full of truth.

We all need the religious attitude toward mystery to help us instinctively remember we are a part of God's plan and each of us is more than the material composition of our bodies. We are all seekers after the image and likeness of God. Prime examples of mystery are the mystery of God, the mystery of life, and the mystery of ourselves. But with equal importance, we need to hold on to the scientific attitude toward paradox (in the same way St. Paul recommended we must all test the spirit, as found in 1 John 4:1 and Romans 12:2), making certain we have pushed the boundaries of truth to their fullest extent. Without this relentless testing of the scientific method, it is too easy for us to fall into unnecessary complications and complexities.

---

[113] Antoine de Saint Exupery, *The Little Prince* (New York: Harcourt, Brace and World Inc., 1943), 93.

With great wisdom, Albert Einstein once said the best solution to any problem is always the simplest solution—but not too simple. Consider the struggle that has been going on for decades inside physics to find a way of uniting the universe of the very large with the universe of the very small. This paradox can only be solved by combining general relativity (the laws of the physics of the very large) with quantum mechanics (the laws of the physics of the very small) and in the process uniting the four fundamental forces (gravity, electromagnetic, weak nuclear force, and strong nuclear force). This holy grail of physics would hold the possibility for science to find a single unified physical theory. At the moment, a successful solution to this paradox is still a long way off.

As an example of how far science can go to resolve a paradox, let us take a little time to understand how much progress science has made in all of its attempts to interpret quantum mechanics. The orthodox interpretation of quantum mechanics is called the Copenhagen interpretation after the location of the physics institute founded by the Copenhagen interpretation's chief architect, Neils Bohr. Today the majority of the world's scientists accept the Copenhagen interpretation as truth. According to the Copenhagen interpretation (see chapter 2), all subatomic particles have momentum and energy that may be represented by an un-measurable mathematical wave function. This wave function may, or more likely may not, have any objective reality.

The quantum mechanical wave function has the ability to be in many places at once. It may penetrate barriers, and it may come into a relationship with itself in what is called a quantum superposition. Human beings as observers are blissfully unaware of the presence of this wave function until a measurement has been made in an experiment crafted by human beings. The outcome of the measurement, according to Copenhagen orthodoxy, is the bestowing upon a subatomic particle measured values of energy, timing, position, momentum, and spin but not all of its properties at the same time. At the moment of measurement, the particle's wave function collapses (its wave function instantaneously ceases to exist, according to Copenhagen) and reality is bestowed upon the particle, along with its measurable properties. (In this regard think about the moment from the movie *The Wizard of Oz* when the scarecrow receives his PhT, "Dr. of Thinkology" degree from the wizard.) Only some but not all of its properties are measurable.

A set of rules called the Heisenberg uncertainty principles defines the degree to which any of the particle's properties may be accurately measured to the detriment of its other properties, which remain unmeasured. According to Copenhagen, reality is only bestowed upon matter and energy when a humanly designed and executed measurement is conducted, and only the measureable properties of the particle (i.e., in this particular experiment) are elevated to the level of reality. Therefore, according to Copenhagen, the un-measureable properties of a particle (for a particular experiment) literally *do not exist.*

Taken literally, as most of today's scientists do, Copenhagen implies there is no reality without human observation, which is to say that human observation is the ultimate creator of human reality! Put another way, until a human being has experienced something, this something simply doesn't exist for human beings. The implication is that independent of human observation, there is no such thing as objective reality.

The Copenhagen interpretation deeply troubled many scientists because of the brazenness of its conclusion. One such troubled scientist was Hugh Everett who at the time (the 1950s) was a graduate student working under John Archibald Wheeler at Princeton. Everett couldn't stomach Copenhagen's insistence that a something-less-than-real wave function must completely collapse at the time of measurement. Everett's PhD thesis contained the development of an alternative interpretation of quantum mechanics, which has come to be called the many worlds interpretation.[114]

Everett reasoned that when a measurement takes place, the particle's wave function continues on as the superposition of all possible outcomes of the experiments. However, each possible outcome defines its own new

---

[114]   The many worlds interpretation of quantum mechanics was developed by Hugh Everett in the 1950s as an alternative to the Copenhagen interpretation, which Everett found unacceptable because of its reliance on wave function collapse at the moment of measurement. This interpretation holds that at the time of measurement, only one of the many possible measurements is made here on Earth, but all of the other unselected measurement possibilities are played out in other universes. Neils Bohr didn't like these ideas, so John Archibald Wheeler, who was Hugh Everett's PhD thesis advisor, asked Everett to tone down some of the more objectionable aspects of his work. However, today there is renewed interest in this alternative interpretation of quantum mechanics.

world (i.e., universe). The measurement recorded by the experiment is simply one of the many possible outcomes that happen to take place in our world. All of the other possible outcomes, according to Everett, are then associated with a quantum superposition of the particle's many possible futures, each being measured in one of the many new worlds being generated to support the measurement.

Unlike Copenhagen, which puts us humans in a position of creating reality, the many world interpretation puts humans in a position of creating the composition of new universes (and without limit!). Both the Copenhagen interpretation and the many worlds interpretation are bestowing on human beings the complementary God-like powers of reflecting reality (Copenhagen) or creating the order of new universes (many worlds). In both cases, powers beyond human ability are being placed in the hands of human beings. This unacceptable situation is *hubris*. It cries out for a new interpretation of quantum mechanics that corrects this impossible status quo and restores us humans to our real status as seekers of truth.

Based on our work, we feel the solution lies with defining a new interpretation of quantum mechanics that includes the actions of God. Therefore, as we have stated earlier, we propose (as we described in chapter 4) that the meaning of quantum mechanics is derived solely from a recognition that God, who is *not* a part of our material universe, communicates with our material universe via quantum mechanics in a way that does not violate our universe's rules of the road (i.e., the scientific laws created by God).

In the Bible (1 John 4:1) we are told that God comes to us in the flesh, although God is clearly from a different place altogether. Is it this coming in the flesh that corresponds to God's communications with our material universe?

CHAPTER 10

IT IS SPIRIT THAT BRINGS GOD'S VIRTUES INTO OUR LIVES

Jesus referred to Satan, the evil one, as the prince (or perhaps the principle) of this world. How true this statement is. The genetic coding that promotes our survival emotions (which are entirely of this material world) is comprised of a complex molecular pattern that is the product of God's natural selection process that was discovered by Darwin. Natural selection is driven by survival of the fittest, in which a species is refined and tested over many generations. Each new generation with its modifications (mutations) is subjected to changing environmental conditions. Human survival emotions represent the collective biological wisdom of how human beings are to survive in the harsh environment of this world.

When Jesus teaches us to love our enemies (pointing out that even the outsiders love their friends and family), he challenges us to look beyond our survival instincts and emotions and seek God's ways instead of our own. God is calling us to a higher plane of virtuous being. Godly virtues are communicated directly to our bodies through the complex interactions within our central nervous system. They occur with the simultaneous collapse of many quantum wave functions during God's communications.

God's information is being processed within our brains through the exchange of electrical and chemical signals that operate the switching nerve cells called synapses. Our brains contain tens of billions of synapses, each one being a potential receptor for bits of the divine information in the form of collapsing chemical ions and electronic wave functions. As it is written inside our brains, this godly information becomes irreversibly transferred onto the complex web of our body's central nervous system.

We may all experience certain survival emotions, such as anger and fear, in each day of our lives. However, for most of us (perhaps all of us), a counter experience is generated deep within us that we may call the spirit of divine virtue. The spirit of virtue offers us a godly alternative to our body's survival emotions, potentially transforming each of us into more godly beings if only we will let it happen. But the choice is always ours.

Some examples of virtue are courage, which leads us out of our fears, compassion, which leads us out of hardness of heart, acceptance, which leads us out of jealousy, mercy, which leads us away from vengeance, humility, which helps us to overcome our pride, and hope, which helps us to rise above despair. God leads us and teaches us by instilling these virtues. But accepting virtue is always a choice, an exercise of the free will in each of us. God never compels us to accept virtue; if God did, we would then be nothing more than programmed robots. Some people call this process our conscience. Conscience can be described as that small voice within our heart that always points toward God.

Choosing to live our lives in harmony with the spirit of virtue is a little like mountain climbing. There are many paths leading up the side of a mountain, and many climbers may be making an ascent in their own way. Each climber's path is marked by the ropes with which he or she pulls him or herself upward toward the summit. Occasionally a climber might slip and fall backward down the mountain. At these times of crisis, other climbers must reach out and offer a helping hand to falling climber to help him or her regain a sure footing. Once secure, the climber will continue the ascent and perhaps return the favor by helping others who are struggling to continue their own upward climb. As the summit is approached, the distinction between the paths of the individual climbers becomes less clear as their ropes come closer and closer together. At the summit, all paths seem to melt together into the shining, pure white light of God's presence.

Our sense of virtue is constantly being refreshed by God. Divine virtue is the cornerstone of what we call our humanity. Consider how we commonly use the term humanity and all of its derivative words. Various human societies promote the virtues of kindness and compassion toward their fellow men, women, and creatures. These humanitarian organizations (for instance the Red Cross and the SPCA) rely on human caring, compassion, and respect to reach out and help people and fellow creatures around the world who have fallen victims to natural disasters, wars, and illness. Though expressed in many different ways and acted out on many venues, human virtue has always been and always will be the truest mark and highest expression of our humanity. This is because it is virtue, in all its forms, that ties us securely and forever to the goodness of God! When God speaks to us, it is like we are God's students. Like all students, we have suddenly reached the "aha" moment in God's lecture when we have internalized a godly virtue.

CHAPTER 11

PASSOVER AND EASTER

The book of Exodus in the Bible tells the story of Israel's liberation from slavery in Egypt. When the Israelites of the Old Testament were freed from the slavery, Moses led them around in the wilderness (today's Sinai desert) for the next forty years. The forty years had a special significance because this was the time required for the generation born into slavery to die off so the freeborn men and women (born during the forty years of wandering) of the next generation could enter the Promised Land.

Many times during those forty years, the slave-born people longed for the relative comforts of their lives back in Egypt. Some were even willing to trade their new freedom for the flesh pots of Egypt. This is thought to refer to ceramic jars containing preserved meat. The Jewish holiday of Passover celebrates how God fought for the Israelites to secure their freedom from slavery. Although Moses was very persuasive, it took a long, hard fought battle to convince Pharaoh to let the Israelites go. In the end, only God's parting (and refilling) of the Red Sea convinced the Egyptians not to follow Israel any further, giving up forever their intention to reenslave Israel. Freedom, it seems, was the most precious gift God could bestow on his people, and more than anything else in life, freedom was and is worth fighting for. Consider the American revolution of 1776 and the American civil war of the 1860s. Consider Egyptian and Libyan revolutions of 2011. Freedom is one of the very few causes that might actually be worth dying for.

Christians believe Jesus died on the cross to procure forgiveness for us and liberated us from the slavery of our sins, much in the same way, by analogy, as Israel was freed from Egypt. What does this mean? Did Jesus literally have to die to secure God's forgiveness for the sins to be committed by people everywhere in all future generations? Was his death the price that had to be paid to secure the release of human beings from their heritage of a sinful nature? How was this accomplished? Was there some kind of an accounting system that totaled up all of humanity's past, present, and future sins, and balanced them against the pain and anguish suffered by Jesus as he died on the cross? This doesn't seem like a very useful way to understand God's plan for humanity, so let us consider some alternative understandings.

In the Bible story from the book of Exodus, God fought for Israel's freedom from slavery. Christians believe Jesus died on the cross for the same purpose: freedom from the slavery of sin. In the Jewish tradition, people carrying a heavy load of guilt would go to the temple and ask the priest to sacrifice an animal on their behalf, thus expunging the guilt of their sins. In the book of Genesis, Abraham was commanded by God to sacrifice his son, Isaac. As Abraham prepared to kill Isaac for the sacrifice, God sent a ram as a proxy sacrifice in Isaac's stead. In the same way, Jesus's death on the cross may be understood as a proxy sacrifice for us all. In one of the greatest of all religious mysteries, God (in the form of the man Jesus) sacrificed himself to expunge the *guilt* of all humanity!

Let us restate this important point: our human survival emotions are hardwired into our DNA as a direct result of countless generations of our evolution. These emotions are necessary for our personal survival during times of extreme need, but when expressed to excess during less-stressful periods, they become the source of great personal trouble, pain, and decay. A life habitually indulging in excessive and unreasonable fear, anger, hate, lust, greed, or pride will be a life that is out of control and locked into a dangerous downward spiral. Such a life is like that portrayed by Moses in Deuteronomy 30:19. He described it as the life of a person who has chosen the path of death as opposed to the path of life Moses recommends. Such people have allowed themselves to become enslaved to their own emotions. Perhaps in the beginning these emotional indulgences felt good. However, like the addictive substances they really are, such indulgences became harder and harder to resist. Eventually they take over lives. People who are held captive by their own emotions are in need of rescue.

St. Paul warned in Romans 6:12-23, "Not let sin exercise dominion in your mortal bodies, to make you obey their passions." Paul goes on to say, "No longer present your members to sin as instruments of wickedness, but present yourselves to God as those who have been brought from death to life, and present your members to God as instruments of righteousness." Finally Paul observes, "If you present yourself to anyone as obedient slaves, you are slaves of the one whom you obey, either sin, which leads to death, or of obedience [to God] which leads to righteousness. But thanks to God that you, once having been slaves of sin, have become obedient from the heart ..."

Herein resides the basis of all ethics. We all must confront the choice either to live a life of addiction to our unbridled survival emotions or ask to receive God's grace in the form of the spirit of virtue. The latter is the path of life, and the former is the path of death. The choice is always in our hands. That is the truth.

God wants no sacrifice. It is God's amazing gift to offer up himself, knowing that humanity requires that once-and-for-all convincing sacrifice.[115]

From the historical perspective, just as in Moses's time God fought to secure the freedom of the Israelites, so in Jesus's time God in the person of Jesus offered himself and struggled for us against the principles of this world to secure our freedom. Today God will fight for us to secure our personal freedom.

Jesus was innocent of all the charges that were being brought against him. Why was he really being tried and executed? The Jewish religious authorities and the Roman civil authorities were scared to death of what might happen to their secure positions (and maybe their lives) if Jesus's rising popularity among the people was not checked and snuffed out immediately. As a vivid example of what they might have feared, consider what we are witnessing today in the same region during massive street protests of the Middle Eastern revolution of the spring of 2011.

Jewish religious law and the Roman civil law were both in danger of becoming the targets of a people's revolt (not unlike Egypt in February 2011!). Perhaps Judas, sensing the nearness of revolution, became disenchanted with Jesus, who he may have felt was not seizing the opportunity of the moment to lead the revolution. It was really the authorities' own churning fear, envy, pride, anger, and even hatred of Jesus that drove them to convict him—not because of anything that he did or did not say.

---

[115] Thomas Troward, *Bible Mystery and Meaning* (Marina Del Rey, CA: Devorss and Company, 1992), 118.

The gospel accounts tell us that Jesus said very little to the authorities during his interrogation. By standing up to the authorities and accepting the necessity of what was to come, Jesus faced down the authorities, forcing them to either accept the validity of who he was and what he was preaching or have him killed. They chose to kill him. For these men, emotion had triumphed over the truth and the right. The one who rose on Easter morning was God, in Christ, who is always with us, always ready to strengthen us in our search for truth.

CHAPTER 12

SACRIFICE

Sacrifice is a principle of life acknowledged by Christianity and by nearly all religious paths. Sacrifice has played an important part in the faith's spiritual understanding since the very beginning. In particular, Jesus's sacrifice on the cross is of profound significance to all Christians. But what was the meaning of Jesus's sacrifice, and what understanding of our own lives can we derive from Jesus's sacrificial act?

Let us first consider the meaning of sacrifice in the context of those cycles obeyed by all life. All living things, including ourselves, receive their energy for living by eating other living things. The only major exceptions to this rule are drinking water and the process of photosynthesis, by which many plants turn sunlight into a form of energy that fuels their growth. It is only within living (or formerly living) tissues of living things that we find the kinds of proteins, carbohydrates, and amino acids our bodies require to build and sustain life. Therefore, each of our lives is sustained by eating some other form of life that has laid down its life as a *sacrifice* to sustain ours.

All living things derive their lives from the sacrifices made by other living things. In the end, all living things are required to ultimately make similar sacrifices and thereby sustain the lives of others. What Jesus did by making the ultimate sacrifice on the cross was to feed each of us by freely sacrificing his human body so we may partake of his spiritual body.

All churches celebrate Jesus's sacrifice with a ritual meal called the Eucharist. By allowing himself to be sacrificed, Jesus teaches us that life is only sustained by the constant sacrifices of others. This is how love is expressed by all who live. God, in Jesus, expressed his divine love for humanity by making the ultimate sacrifice on the cross. Sacrifice is the giving of a life rather than the taking of a life. Of course, since Jesus was God incarnate, God in Jesus did not cease to exist on the cross but rose from the dead man Jesus on Easter morning, expressing his love for all. God's loving sacrifice gives us all a hope in a future where no matter what has happened, the God of love and truth will always be there for us.

Evolution and Sacrifice

Science teaches us about the profound importance of sacrifice by way of the theory of evolution. Darwin's evolutionary theory teaches that any species will advance and change only through the creation of certain special individuals. These rare individuals who have received a successful mutation are able (by virtue of the sacrifice of others) to live longer lives, enabling them to mate more often and pass along their successfully modified genetic code to the future generations of his or her species.

It is this new genetic code that offers hope for the future of the species, as new individuals possessing this genetic code confront the uncertain future of a changing environment. However, these genetically successful individual are only able to live longer and pass along their new genes through the sacrifices of others (coupled with the driving force of the successful individual's survival instincts). Countless other individuals (of various species) are required to lay down their lives as food for the genetically successful individual so his or her newly mutated genes will bring the hope of a successful future to his or her species.

Reflecting on sacrifice and on the biological cycles of life, we must consider the possibility that we humans may be part of a parallel spiritual evolutionary process. Perhaps it is by a process of spiritual evolution that people are shaken out of their fixations on the emotions of personal survival and brought to see that their lives (as Jesus showed us with his words and his example) are truly a composite of personal sacrifice and transformation. It is the teamwork of these two forces that moves humanity step by step closer to God. All human lives become an ongoing meshing of the biological cycle of life and the spiritual cycles of sacrifice and enlightenment. We live, eat, thrive, procreate, think, feel, pray, meditate, and receive the insights that give birth to our personal hopes and dreams. However, all along the way, we make sacrifices. As with all life, we act out a Eucharistic dance that is comprised in equal measure of sacrifice and enlightenment.

Let us reflect further on the Eucharistic dance from a related but different direction. As we have already seen in chapter 4, each of us acts as a creator and source of information as we walk along life's journey. Each bit of information we create requires the investment of a small amount of energy, making these bits distinguishable against the background of our environment. For instance, if we are signaling with a light source in such a way that each bit is represented by the presence or absence of light against the background of the sky, it is very important to be sure the light source is of stronger intensity than the sky's ambient light. Naturally more light energy is required to signal at noon than at midnight. It takes energy to lift a pen and write down words on a piece of paper. It takes energy for us to force our fingers to press the keys of a computer keyboard as we enter information into a computer. Likewise it takes energy to make the transistors within a computer switch between the two electronic states, symbolizing the writing and erasing of bits of information.

What happens to all this energy? The answer to this question lies deep within the profoundness of the second law of thermodynamics. The second law tells us that no matter what kind of action is taken when writing a single bit of information, the outcome (in due course) will be an increase in the disorder of the writing energy. In this way the writing energy associated with the bits of information we encode can never be retrieved once it has been returned to the environment. This energy becomes like drop of water falling into an ocean where it can never again be distinguished from the ocean's waters.

The wisdom of the second law of thermodynamics is perfectly describable by the nursery rhyme of Humpty Dumpy. As we all can recall from childhood, "All the king's horses and all the king's men couldn't put Humpty Dumpy back together again." Just as poor Humpty's pieces were too scattered to be located and reconnected, the energy sustaining one bit of information is irretrievably scattered as it is distributed to its surrounding thermal environment.

You might say this energy is lost just as a drop of water is "lost" when it is delivered into an ocean. However, the first law of thermodynamics (conservation of energy) assures us that *none* of the bit's energy is ever truly lost, but what is lost in this transaction is the order represented by the organization of energy into a single bit of information. The bit of information, and the order it represents, is, like Humpty Dumpy, ultimately lost forever. It is this process of information creation and destruction that consumes all information, which must ultimately be sacrificed in order to make room for the new information that is always being created. Spirit is constantly creating new information and renewing the old that is being sacrificed. God assures us of this process, for the Bible tells us (1 Cor. 15:40-49) that all things are being made new. Form (and information from which form is composed) is irretrievably lost as its energy returns to a state of chaos. But it is the spirit that is constantly engaged in the process of renewal.

In physics, the information that is lost by making these sacrifices (as required by the second law) is called entropy. Entropy is equivalent to the amount of disorder created as information is annihilated. Entropy is measured in units of lost information (i.e., bits). A way to think about entropy is to compare a brand-new car with a very old but similar kind of car lying in a junkyard. When a car goes to the junkyard, in all likelihood it still has all of the same essential parts that it had as a new car. The change from new to old is that the parts of the old car have now worn out. The information that was once contained in the precision fitting of the car's parts (consistent with the manufacturer's specifications) has now been hopelessly lost. The old car no longer has any value other than the value of its scrap metal (which is small compared to the value of the car when it was new).

The real value of the car was contained in the information related to the precise fitting of the car's parts (i.e., as specified in the car's design). Very little of the car's value resides in its materials. After a car is worn out, its piston rings will no longer seal the engine's piston against the cylinder walls. Fluid-carrying hoses will crack and no longer be capable of holding high-pressure liquids and gases without leaking. The engine's bearings no longer have precise metal-to-metal contact surfaces that allow them to function with nearly frictionless movement.

Christopher Wren's,[116] the famous eighteenth-century London architect, first job while rebuilding St. Paul's Cathedral after the great fire of London in 1666 was to completely destroy and remove the burned-out shell of the cathedral that stood on the site. The old must always be sacrificed to make way for the new. New trees in a forest can only find their places in the sun by the sacrifice of the surrounding older trees that have at last died and fallen to the ground. Our own bodies are a

---

[116] Christopher Wren (1632–1723) was an English architect and developer. Wren almost singlehandedly rebuilt London after the great fire of 1666.

prime example of this principle. The acceptance of life's fleeting nature is really a philosophic statement of the second law of thermodynamics. Understanding this principle is a very important step for all of us in the process of acquiring personal wisdom.

The principle of sacrifice applies equally well to all things, from the smallest molecule to the universe itself. In the final analysis, the entire universe will continue to grow ever larger as the expansion that started in the big bang continues. All actions and processes, both natural and manmade, taking place within our universe are played out over the course of history. No energy is ever lost, but as a result of the entropy process, this energy is finely scattered and distributed in a way that the information it once contained is lost forever (In its final chapter, our universe will become so large that all of its matter and energy will be spread so thinly that sufficient energy can no longer be collected in any one place to create a new bit of information.)

In these final moments (or perhaps eons), our universe will slip into a totally static state in which nothing new ever happens or *can ever happen*. Time, the ultimate measure of change, will cease to exist. Science calls our universe's ultimate fate the heat death and regards it as that point in the history of the universe when further activity of any kind is not possible. Quite literally, at the heat death, our universe will have become a burned-out cinder.

Many people find the contemplation of our universe's fate to be a very depressing experience. However, we think it is important to remember that the heat death is nothing more than the combined record of all of the changes and sacrifices that have occurred within our universe since its birth in the big bang. In this regard, it is instructive to remember that the unit of measure of entropy is the same as that of information. When the universe finally slips into its heat death, the sum total of all its entropy will be exactly equal to all of the information that was ever created within our universe since the singularity at the moment of the big bang. Jesus taught us by his own example that sacrifice is love. Therefore, the heat death itself is nothing less than a final record of all the love that has been given and received within our universe, from the moment of the big bang until the heat death throughout the vast eons of existence.

Another conclusion we must draw from these ruminations is that our universe, like ourselves, has a beginning and an ending (from alpha to omega). Nothing in or about our universe is eternal. Only God is eternal.

Our hope always lies in drawing closer to God, who in the beginning hovered over the deep.

Eternal Life: A Student-Teacher Dialogue

Student to teacher: What is eternal life? The psalmist (Psalm 90:10) says our life lasts seventy years—eighty perhaps, by reason of strength. The meaning of "eternal" must be the chain of life that links generation to generation. But for the individual death is certain, is it not?

You are spirit, says the teacher. Spirit manifests on earth as your human body with its human mind. You are child of the parent—child-spirit begotten of, maintained, and educated by your Parent-Spirit in this earth-time and experience.

Spirit first—the body is the Spirit's manifestation. All around you the phenomena of the earth and the universe manifest the same Parent-Spirit. All are interrelated in the Spirit's love.

The Spirit has ordained that we phenomena on earth are each other's nourishment.

All of us feed on the Parent-Spirit, who feeds us all who feed on each other. That is how we all are kept in life, now and forever.

If in faith I am given to comprehend that I am spirit, I will strive to make that understanding the point from where henceforth I view the daily experience of life—life on earth within the universe, life of spirit in the Spirit's eternal presence.

But the man of science would want to know the mechanics of resurrection and life. When my body quits functioning, it is transformed in death from its human form and reduced to its mineral and chemical component: "ashes."

The body of Christ, however, is transformed into Spirit. Jesus left no ashes behind but instead left the imprimatur of a human face on a linen cloth—the face of the human man.

Solemn witness—eloquent in silence. Portrait of love that has overcome death.

No more ashes, but instead
Spirit-Life
and the invitation to follow fearlessly.

What Will We Do When the Oil Runs Out?

Since the beginning of the industrial revolution in the early nineteenth century, petroleum-based fuels have played a key role in supplying the energy of progress. Granted, steam engines were the principle movers of the early years of the industrial revolution, and steam engines largely ran on coal for fuel. However, the need for oil lubrications, kerosene for lighting, and later bunker oil and gasoline as fuels for internal combustion engines caused petroleum-based fuels to assume a position of paramount importance within the worldwide developing industrial economies of the late nineteenth and early twentieth centuries.

All through this period, a race was on to discover new sources of petroleum. Nature cooperated in this search, and rich oil fields in Pennsylvania, Texas, Oklahoma, and California opened up, fueling the expanding industries of the United States, the United Kingdom, and other countries in Western Europe and East Asia. In fact, without these expanding petroleum resources coming online in the late nineteenth and early twentieth centuries, the process of world industrialization would not have proceeded at anywhere near the rapid rate it did. It is not too strong a statement to say that the entire process of world industrialization was built on readily available, cheap petroleum fuels.

However, none of the earth's resources are available without limit, and petroleum is no exception. As petroleum resources within the United States have become depleted, the search for new sources of oil has been extended throughout the entire world. Major petroleum resources were discovered in the Middle East, Canada, Latin America, the North Sea area, and Russia. However, while vast, these resources are not without limit.

In a kind of Malthusian process, as more countries became industrialized (i.e., China and India today) and money started to flow into the hands of industrial workers (who formerly were poor agricultural workers), these people spent their new wealth on the luxuries they coveted while watching the citizens of older industrial countries. High on the shopping lists of these new industrial workers were fine foods (with higher meat content), nice clothes, world travel vacations, and *cars*!

A common element on all of these wish lists is petroleum. Without petroleum, there is no corn to fatten the beef cattle, no energy to produce the clothing, no jet fuel to fly tourists across oceans, and no gasoline to fill the empty gas tanks of new cars in developing countries. Industrial expansion in new parts of the world will sharply increase the worldwide demand for petroleum. What remains unknown at present is how much longer the earth's petroleum resources can continue to meet this increasing worldwide demand. If this process continues unchecked, there must come a time when the worldwide demand for petroleum will no longer be satisfied by the existing known petroleum resources.

Many experts now believe that roughly half the world's total supply of petroleum has already been pumped and consumed. If this is true, the world is now facing the hard reality of how to continue the process of world industrialization on a petroleum supply that is already half gone! The upshot of this situation is a near-term prospect of steeply rising petroleum prices and frequent shortages.

Many people look to new and alternative fuel sources that hopefully will become available soon. Let us hope so, but what will it be? Nuclear power (with all of its safety risks) is probably the best near-term prospect that is not horribly polluting. Coal is the other near-term prospect, but a check on China's air quality will answer the question of what will be sacrificed if coal were to replace petroleum as the world's dominant fuel.

However, both nuclear power and coal produce electricity, an energy source that is not ideally suited for transportation. Potentially electricity could power most surface transportation, but to facilitate such a change will require a giant leap forward in battery technology. Other alternative fuels, such as ethanol and biomass, could potentially replace petroleum, but these alternatives are a long way off from practical implementation. Wind, tide, geothermal, and photovoltaic energy sources are also possibilities, but they are also producers of electricity and are unlikely to generate the same volume of energy as nuclear energy or coal. Facing this uncertain energy future, the world must learn to pursue industrialization and commerce in a way that is more reliant on the electronic data processing, and less reliant on a physical transfer of goods and services.

Perhaps energy resource necessity will drive the world in this direction and into a kind of postindustrial business alternative. Already significant commerce is conducted over the Internet. In addition to commerce on the Internet, many people today engage in worldwide computer gaming. These games have risen to such a high level of reality (such as World of

Warcraft[117]) that players sometimes lose track of which world they are most at home in—the everyday, natural world of trees, dogs, and roads (and the laws of physics, chemistry, and biology) or the created world of their Internet-based computer games where reality has been defined by the game's creator, who is a human being.

It is not hard to imagine a future in which many people's primary job is performed somewhere within the cyberspace of the Internet. More and more this virtual space will come to define reality for these people, and the natural world of plants, trees, rocks, and oceans (and the laws of science) will progressively hold less and less significance to many, if not most, of the world's people. However, two important questions emerge for this futuristic view: first, how will all the electricity be generated to power this emerging and vastly expanded worldwide cyber-network, and second, and perhaps most importantly, who will produce the food needed to feed these cyber workers?

According to Moore's law,[118] the amount of computing capability that can be fabricated on a fixed area of a semiconductor will double about every eighteen months. Therefore, as the amount of computing horsepower increases exponentially over time, the cost of computing (per bit) is decreasing exponentially. However, what is not decreasing but is in fact rapidly increasing is the electrical energy needed to power this skyrocketing worldwide computing ability. As an example of how this trend is already starting today, it is very interesting to note that both Google and Yahoo are building massive server farms on the banks of the Columbia River (in the state of Oregon). The reason is to take advantage of the relatively plentiful and cheap hydroelectric power available in this location.

Today there is a great deal of interest in conservation and the environment. However, what will happen in a future world where people live their lives more in cyberspace than in the natural world? What conflicts will arise over the control of electrical power, food, and perhaps

---

[117] World of Warcraft is a fantasy computer game requiring much preparation to ensure that a player's "avatar" is properly equipped with supplies and weapons to successfully wage virtual war.

[118] Moore's law was discovered by Gordon Moore (born 1929) in the 1960s. Moore's law states the number of transistors on a single IC (i.e., computer chip) will double every eighteen months as a direct result of technological progress. After five decades of progress in semiconductors, Moore's law still holds true.

most importantly water? Will there be struggles (warfare?) between those who live in the natural world and control its resources and those who live in the cyber world and are making all that has come to be recognized in this time as progress?

This is beginning to sound a little like the plot of a science fiction novel, but it is not inconceivable that such a future could become a reality. This conclusion leads to the view of a future world that is vastly different from the world we inhabit today. Since cyber reality is where most people in this future world will really live, there will no longer be any priority placed on environmental concerns, and all of the natural world may eventually be sacrificed to produce food, water, and the all-important electrical power to run computers.

People may choose to live underground to escape the miserable conditions on the earth's surface. These underground colonies might be shared with giant server farms and automated computer hardware fabrication facilities. The dirt and grime of the earth's surface might be given over totally to the raising of food (under very questionable health conditions) and of course the all-important generation of electricity.

A future such as this that is focused entirely on living in a created world rather than in a natural world is very difficult for most of us to contemplate. Even more difficult to conceive of is how such a future could possibly be consistent with God's ultimate plan for humanity. It is hard to escape the conclusion that a humanly created cyberspace life that dominates the life and times of humanity is moving away from rather than toward God. Clearly such a human-created world would no longer require the agency of a creator other than human beings themselves.

The Blue Baseball

In 1970 Apollo 13[119] was moving along its planned trajectory toward a landing on the moon when the unthinkable happened. An oxygen tank on the outside of the command module exploded, causing many of the spacecraft's systems to fail (or become of reduced usefulness), including the vitally important electrical power, communications, and life support systems. The crew was forced to retreat into the lunar landing module, using it as a kind of space lifeboat. The planned lunar landing was cancelled, and Apollo 13's safe return to Earth became the paramount concern of mission control. Since there was no way to turn Apollo 13 around, the only safe way for the craft and its astronauts to return to Earth was to go all the way out to the moon, and then, like a boomerang, be flung back to Earth by the moon's own gravity. The round trip would take four long, frightening days.

As Apollo 13 neared the moon, Earth was quickly disappearing in the crew's rearview mirror. After a very low-altitude half orbit around the moon, Apollo 13 emerged from behind the moon's far side, and there was Earth, our fragile island home, appearing as a small blue baseball! We can only imagine how lonely, sad, frightened, and full of longing for their home the astronauts of Apollo 13 must have felt at this moment. There in the vastness of space, only Earth, our mother, offered them safety, comfort, and life itself, although to the astronauts she now had become only a small blue baseball.

There was no other home for the Apollo 13 astronauts, and in truth there is no other home for any of us. Earth is a very small, fragile and finite place, but it is still our home and the only home we have. God has given us Earth, and we have no reason to think there is any way of finding an alternative home within practical traveling distance (based on the laws of science and how they relate to space-travel technology). We must all respect and care for Earth, our mother. Every day she gives us life, and without her we are surely lost. Scientific and religious prophets alike must unite to seek and find a sense of profound respect and gratefulness for Earth, our mother, and for the one who made her. Without Earth our mother and God our father, we are all surely lost. We must pray to God

---

[119] The Apollo program was NASA's program in the 1960s to land humans on the moon

for the wisdom to make the best possible use of our technological gifts, but we must also have the long-range view to pray for God to send us the seeds of mutations that will move humanity forward toward a future of love, compassion, and peace.

CHAPTER 13

THE EXPANDING EXPRESSIONS
OF EXPERIENCE

Change

What constantly changes but always remains the same? The answer is, of course, change itself. But what are the agents of change? As we have already seen, quantum mechanical interactions assure a steady flow of uncertainty that manifests change, even if nothing else is changing. When you boil it all down to its basic causes, change is composed of a partnership of individual free wills and new ideas that individuals are free to choose to act upon or not act upon. We call these new ideas seeds, and we call the free-will decisions of individuals simply will.

We do not mean to imply that will is solely the act of human beings. Will can be thought of as any act on the part of life. But even beyond living, will, as it is understood in chaos theory, becomes a propensity on the part of all matter and energy to evolve in somewhat predictable but often unexpected ways. The classic example that is given for this phenomenon is a butterfly whose beating wings in Brazil cause the formation of a growing atmospheric disturbance that eventually becomes a hurricane.

However, there is another side to these acts of will, and this side is what we might call the seed. A seed provides fresh new information that has only just become known to an individual. When we speak of a seed, we are thinking of quantum mechanical information that has just been captured at a wave function's collapse as a particle assumes the classical properties of position, momentum, time, and energy.

However, speaking a little more symbolically, we are also thinking of the image of God in Genesis 2 where God is presented as a gardener who is planting the seeds of God's own creativity. Jesus often spoke in his parables about seeds, planting, and reaping. The whole idea of seeds is tightly bound up within Scripture to our experience of God. An individual who encounters a seed may decide to perform an act of will based on information contained in the seed, bringing change. However, an individual may decide to ignore the seed, in which case no change occurs.

It is the marriage of the encounter of will and seed that gives birth to all change. Such encounters may be as basic as a single atom encountering a photon (raising one of its electrons to an elevated state of energy), or it may be as complex as a surgeon who must make a life-or-death decision about how to proceed with a high-risk operation. The common thread in all cases is that an individual commits an act that generates some amount of information that records the results of the individual's decision. It is the combined informational records of many individuals that chart the course of progress as the wealth of human (and other life) experience expands.

But is it possible that something of a general nature can be perceived in this collective experience we call change? Does there exist some kind of template that describes the way in which a population of individuals changes and grows as their experience expands?

Let's consider some examples. The first example is the universe itself. As understood by the science of cosmology, our universe began approximately 13.5 billion years ago at an infinitely small point called the singularity. The singularity contained the exact quantity of matter and energy that will ever exist within our universe. From its birth at the singularity, our universe has grown outward into what it has become today. (The diameter of the present universe is in excess of 25 billion light years!)

The seed of the universe's genesis is the singularity itself. There is really nothing more to understand about the singularity, since it is unlikely the laws of physics ever apply to its contents. Therefore, to the best of our knowledge, there could be no information, no complexity, and no structure within the singularity. Our best guess is that the singularity was uniformly one singularity in the same way that the burning bush encountered by Moses in the wilderness was simply "I AM_____" (you fill in the blank), but the possibilities are limitless. There is not much more that can be said about the singularity other than there was once a singularity and everything that is now a part of our universe was contained within it. We have no idea what it was composed of. The seed of our universe was the singularity, and its will was the outward explosive force that hurtled everything, including space-time itself, into existence.

A second example of an expanding experience is life itself. Life on earth began within the first billion years of the earth's existence (the earth is now 6 to 7 billion years old). In ways that are not yet fully understood, the first single-celled creatures to somehow develop the capability to store their own personal genetic blueprint (and pass it along to the next and future generations) developed spontaneously during this period. As time passed, life continued to change and evolve. However, the way in which genetic information was stored and passed to future generations remained unchanged. Each new expression of life added its own uniqueness to overall experience of life. Life is literally the chronicling of how each creature's existence adds information to the record of life on earth. It is the record itself that is constantly expanding in volume and complexity. The changes evolving life goes through contribute richly to the ever-expanding treasure of genetic information, passing the story of how to live to all succeeding generations.

A third example of expanding experience is the life of an individual. At conception, all of us (plants, animals, humans) are single-celled creatures. Eventually each single-celled individual divides into multiple clusters of copies of that first cell. In time, every new cell that is added to the individual assumes a role that contributes to the formation and growth of all the various bodily organs. Only those genes within a particular cell's DNA that are needed to manufacture the proteins necessary to construct a particular organ are switched on to serve as the blueprint for cell construction. The seed of an individual's life is its DNA, which is determined at conception, and its will is the process of cell division and the switching on of the right genes to build the right proteins to manufacture the right organ.

A fourth example of expanding experience is our Jewish-Christian holy scripture (the Bible). Scripture is not in any way a chronological work. Rather it is an anecdotal record of individual human beings' encounters with God. The seeds of the Bible lie in its accounts of individual encounters with God. Some examples are the "I am" experience Moses had with the burning bush or Jacob's night beside the river when he wrestled with the man (i.e., God) who blessed him, but hurt him. In another time and place, it was the traveler's encounter with the risen Jesus as they walked together along a deserted road.

The Bible is the evolving collection of stories and anecdotes that change in detail but always remain the same in so far as they record encounters of individual human beings with God. Time passed, and the Bible we read today first became written literature when written language was developed (making it possible to write down stories from the oral tradition). Later the now-written stories were copied and recopied by generations of scribes who, in many cases, added and subtracted material based on their own inspiration and wills. The Bible in our time has become a living, breathing body of information constantly in a state of change, challenging us as we read it to determine for ourselves what it really means to us personally. Most people find the meaning of a particular Bible passage changes each time it is read. The events in one's own life and one's own needs have a lot to do with what an individual's heart may hear as the Bible is read.

Is this how change begins in general? Is all of life a dance of seeds and wills that expands the experience of the dancers into forms of ever-increasing complexity as more individuals join the dance? Is it possible to draw any general conclusions about this amazing process?

Here are some general thoughts. Every community starts at a moment of conception. In the life of an individual, this conception is literally the truth. In the case of the universe, the conception was the singularity itself. For life in general, there must have been some moment when a special, giant organic molecule took on a form that allowed it to store genetic information, enabling inheritance and the whole evolutionary tree of life to come to life. Everything has a beginning, and we simply call the beginning conception. But where did conception come from? Can something come from nothing?

It certainly can in quantum mechanics. (Something temporarily produced out of nothing is called a virtual particle, and virtual particles speak to the possibility of godly actions before the big bang!) It is the quantum mechanical nature of initial seeds that starts the whole process rolling, causing conception. Once conception has occurred, there is a beginning time for all communities when a kind of sameness of the component parts exists.

Each living creature begins as a single-celled zygote[120] that in time divides into identical cells as growth begins. Initially there is no structure

---

[120] After conception, the zygote is the first single cell to contain all of the genetic information that is destined to be received by the new individual.

to the community. Almost instantly after the singularity that started our universe, a white-hot cloud of light with no distinguishing features appeared to serve as the starting point for the development of the first generation of stars and galaxies. In all cases, a sameness at conception eventually gives way to an emerging period of what we call differentiation, for lack of a better term. In the differentiation phase, distinguishing features and forms begin to emerge.

With the universe, when sufficient cooling of the primordial fireball happened, particles such as protons and neutrons became condensed out of the white-hot mass to form the atomic nuclei that in time attracted electrons to form atoms. The process of differentiation continued throughout the early phase of the universe that is called the inflationary period. During inflation, the universe expanded rapidly, and as more details began condensing out of the primordial fireball, the formation of the first stars and galaxies began.

With living things, similar things happen. Sometime after a new life's zygote stage, an incredible thing happens in which rapidly dividing cells begin to realize (as the correct genes are switched on) they were intended to be heart cells or eye cells or brain cells. At this realization, those portions of the zygote's DNA are switched on that encode the amino acid requirements needed to manufacture the unique organ-specific proteins required to build a particular organ. At the same time, each living thing goes through a phase of differentiation in which a cascade of the newly divided cells associated with an individual's development begin to acquire a sense of what particular organ each new cell is intended to build. Each cell acts according to its instructions, and organ by organ, development of the entire body is the final result.

With life as a whole, similar events have taken place. Biologists talk about there being a small, warm pond where the first life appeared on the surface of the young earth. Science has tried repeatedly to duplicate the conditions encountered in the small, warm pond envisioned in the minds of scientists and have come up with nothing more closely resembling early life than black-brown gunk.

It turns out that the key to life at the molecular level is not so much in its chemical ingredients as in how these ingredients are put together. All of the correct proportions of elemental oxygen, nitrogen, phosphorous, sulfur, carbon, hydrogen, and sodium can be present in these experiments, and the results will still be black-brown gunk. The trick is how to make the right molecular structure of life naturally synthesize within the reaction vessel. Such a trick has not yet been mastered and may never be mastered.

Many scientists feel the odds are overwhelming against the random synthesis of DNA and other giant organic molecules associated with genetic information and heredity. Only with the synthesis of these giant organic molecules (such as DNA) did it become possible for emerging life to encode its own genetic information, holding the potential to pass genetic information on to future generations. It is this inherited genetic information that makes the process of evolution and natural selection even possible. Once evolution became possible, life was free to develop and change into the myriad of forms that have appeared and disappeared on earth over the last five billion years.

All along the way, as differentiation and diversity are developing, a constantly changing dance takes place in which new seeds are sown and wills are exercised. These seeds come from the spirit, and the decisions are from the wills of individuals. Every decision results in an event that creates a fork in the road along the highway of our unfolding universe. The events of will are chronicled as the information recording their occurrence. Should this massive amount of information be gathered into a single place, the entire story of our unfolding universe, down to the history of each atom, could be told in its entirety.

However, information itself is a physical thing, and it too has a lifetime, just like the material makeup of the events it is recording. Entropy works on all information, dispersing it so completely that in time every message is lost. We might think of the information before the process of entropy begins as being a measure of the system's organization.

As an example, consider all of the things at a child's birthday party that have been carefully organized by the child's parents before the party begins. However, when the party is over—children being children and great agents of entropy—the children have succeeded in scattering all of the party's materials like plates, cups, tableware, napkins, cookies, ice cream, and cakes far and wide. It is the scattering that represents the work of entropy as it causes the disappearance of the information contained in order. It is organization and not matter and energy that are lost to entropy.

It is in organization and order that information resides. For instance, the information contained in three rocks piled on top of each other represents a trail marker (answering the question of which way to proceed down the trail). On the other hand, should these same three rocks be scattered on the ground by storms or animals, they would no longer mean anything to a hiker except appearing as just one more pile of rocks along the trail. Information is lost in the scattering process associated with entropy.

Surely there is something deep within our universe that sows the seeds starting the dance of organization and information. But just as surely as Cinderella's coach must turn back into a pumpkin at midnight, the information recording the universe's organizing events is returned to chaos by the process of entropy. Every dog has his day, but his day doesn't last forever. The wisdom of science and religion are in perfect agreement on this point, and the truth is forever unified!

Faith

Truth absolute; it is. That is a statement of faith because truth cannot be defined or described. We can bring proof only of subjective truth, which is relative and provisional at best (and cause for lawyers' arguments). Nevertheless, our every morning's first conscious moment admonishes us to live life based on the truth that is. People who become aware of that admonition will pledge their every effort to the search for truth to build on that foundation the conduct of their lives on earth.

That pledge is faith. In faith we trust to receive strength and light to guide us through the landscape of daily changing circumstances. Faith keeps us focused on the searcher's path. Faith bids us to rise above doubt and above rational, objective critique, hold judgment in abeyance, and press on to be ready when the light of truth begins to dawn.

Faith is hard work. It requires discipline and courage to walk into the unknown. The paths toward truth are many, and all require faith and dedication. Science and religion are presently the most visible paths. People of the dominant cultures on earth today tend to invest science with the supreme authority to verify the truth. But some of us seek truth along the pathways of religion. Both science and religion, divided even within themselves, are perceived to be separate and in competition against each other as sole authority in matters of truth.

So challenged by reason's critique and tempted with distractions in the marketplaces of things and ideas, faith is the sometimes-patent but often barely discernible guide through the uncertain terrain of provisionary truths. It requires steadfast resolve and patience until all paths of science and religion are brought to light and shown forth united in truth. No human effort can capture and demonstrate truth. Truth alone bestows that revelation in a moment of grace.

The most difficult task of faith is the task of discrimination. What is genuine and what contrived? What is honest, and what is equivocation?

The Gospel of Thomas[121] quotes Jesus saying, "Know what is before you. That which is hidden will be revealed." That is an admonition you can let walk with you to teach you to dismiss nothing out of hand. Hold judgment in abeyance and ask, "Whose spirit's child are you who comes into the circle of my perception?" Over time, the truth of the person or the phenomenon will come to light.

There is a pattern in the teaching and learning path toward the truth. Students attracted to a teacher will follow the teaching easily as long as the subject can be recognized by the norm of beliefs and tradition. But when the teacher takes you into unfamiliar depths where old word associations gain new meaning, then the student's mind comes into crisis.

We can take an episode in St. John's gospel as an example. In the course of the gospel's sixth chapter, Jesus gives a familiar practice—the eating of the community meal—a new meaning. A large number of people had just partaken of a common meal that had been consecrated and provided by Jesus the teacher. That experience was amazing enough in that the teacher did not bring a boatload of supplies to his retreat. There was a depth of meaning to ponder. But when the teacher took the meaning of the experience to yet a higher level, the disciples were shocked. The teacher said, "I Am the bread from heaven. Your forefathers ate manna in the desert and died. When you eat this bread, my flesh, and drink this wine, my blood, you will live." That teaching made the majority of the students shake their heads and walk away.

Fritz writes, "I have been at such points a couple of times. Once in grammar school in Germany, our class was visited by a professor with his student teachers. We gave one student teacher a very hard time. He tried to teach us that two parallel lines drawn to infinity will cross each other. '*No way! Impossible!*' we yelled. We had him all flustered and refused to

---

121   The Gospel of Thomas is a noncanonical gospel that is perhaps Gnostic or proto-Gnostic.

listen to his attempts to clarify those mathematics. His professor left him foundering in our protest."

Every great endeavor comes to such a point of impossibility. At such a point, the student-disciple will find no hold in what he knew before. This is where faith has to keep the truth-seeker's mind open and with the teacher. Many if not most students capitulate and revert to shallow backpedalling: "Who can listen to that? He's crazy. He's not a prophet. No prophet comes from Galilee." These are familiar phrases picked up along the way.

Those who stay with the teacher likely understand just as little of the teaching as the others who leave. But faith and the hope that in time the truth will dawn for them keep them close. The entire Christian faith is centered on this teaching begun at the Sea of Galilee: The body of Christ is the bread of heaven. The blood of Christ is the cup of salvation.

Faith carried those who stayed with the teacher across the abyss of uncertainty. They stayed with him because they knew in whom they had believed. When Jesus asked his closest friends, "Do you want to go too?" Peter answered, "Where shall we go? You have words of eternal life. We have come to believe and know that you are Messiah, the Son of the Most High." Most people would say that such childlike faith is blind. But there is in it a knowing above and beyond the grasp of cerebral logic.

R. M. Rilke,[122] a German poet of the 1920s, portrayed this call to faith in his poem "*Stimme im Dornbusch.*" The poem's last lines portray that childlike discipleship:

> *Der Held ertrotzt es sich auf seine Weise.*
> *Doch andre folgen nur und gehen froh*
> *als gingen sie durch Luefte, durch Porphyr*

*   *   *

> *The hero stubbornly insists on his own way.*
> *But others only follow and walk clear*
> *through porphyry as if through air*

---

[122]   R. M. Rilke (1875–1926) was a German poet.

The religious disciple is not the only one in need of faith. Researchers and their students and research assistants also have to have faith, discipline, and patience. They must have patience not only in the pursuit of truth in their findings but also with knowledgeable challengers and shallow scoffers and ignorant raisers of unjustified hopes (the latter especially in the field of medical research). Discipline is necessary for researchers to stay with the hypothesis they have believed in. They also must have courage to let go of a hypothesis that is proven to be untenable. Faith in scientific research is the conviction that the advance into the not-yet-analyzed residuum of the unknown has been furthered whatever the outcome, and it is useful work.

Humor is another side of faith that lightens the load of stress we carry in our search for truth. Whether it is a Baptist arguing exegesis with a Catholic or Albert Einstein debating Niels Bohr in the fields of quantum uncertainty, all of us can look up, recognize each other as human friends, and start bailing the flood of emotions together that threatens to drown us. We are, after all, in the same boat.

We all need to have faith that someone will have hold of the tiller to keep the sail in the wind and to steer the boat to shore with truth the invisible compass.

May the truth shine on all of us and bless our lives.

CHAPTER 14

THE PRACTICE OF TRUTH

Without practice, a philosophy is of questionable value. How can we apply a united truth to everyday situations in a way that makes all our lives enriched and enlightened? In this penultimate chapter, we explore ways of practice in which the truth of science supports the search for religious truth and ways in which religious truth supports the search for scientific truth.

The Dalai Lama,[123] who is the spiritual leader of the world's Tibetan Buddhists, once said the best religion for each person to follow is the one in which he or she grew up. This is because the religion we learned as children is the religion we have the deepest connection with and that resonates most strongly within the deepest layers of our souls.

We think he is right. It is best for each of us to continue following the spiritual path we each started on as children. We believe there are many paths to God, and the one that is right for each of us is the one to which each of us feels most deeply connected. In the authors' case it is Christianity; for others it might be Hinduism, Buddhism, Taoism, or one of the numerous tribal religious paths of indigenous people worldwide. We think all of us need to find our own spiritual path and stay on it.

That's not to say we cannot benefit from knowledge of other paths. By all means, learn as much as you can about any or all paths to God, for these other paths can only deeply enrich your own personal experience. However, it is best for each of us to continue to practice our own unique, heartfelt path. If you were not raised in a religious tradition, the choice is

---

[123] The Dalai Lama, Tenzin Gyatso, (born 1935) is the spiritual leader of the Buddhist people of Tibet.

yours alone. Remember, all religious traditions are uniquely valid paths to God, and the choice is yours alone.

All practice involves prayer and meditation. In very simple terms, the difference between prayer and meditation is that prayer is talking to God but meditation is listening for God. There is a place for both, so both should be practiced according to the wisdom and tradition of your own personal path. All practice involves relationships with other human beings. This area is very important because it involves us directly making hard choices.

It is very important that each of us take our relationships with other people very seriously. We must treat each other with the utmost of respect and hold each other in the highest regard. With the first hint of disrespectful feelings toward the other person, temptation will see an opportunity to rush into our lives and tempt us to do its bidding. Of course, this is not to say there will never be times when we need to protect ourselves against the ill intentions of others. Of course there will be! But be on your guard against fighting evil with evil. Should this ever be the case, good always loses out and evil gains the upper hand.

Be vigilant to watch for each and every God-given opportunity. God sprinkles our world with the seeds of his plans and purpose. These are hints about God's plan for our lives. The value of an opportunity must be carefully weighted, and the final decision to accept or reject the opportunity is of course yours alone. Nevertheless, opportunities are always the fuel of our lives. Each of us writes the information of our personal sagas based on these opportunities that suddenly arise, often when we least expect them.

Opportunities often open the door to experiences we would not have access to otherwise. Oftentimes opportunity comes in the form of meeting another person who can enrich our life (and we can enrich each other's life) in ways we could not have anticipated. Often the answer to a difficult question walks right up to us in the form of a new person with a fresh point of view on an old and nagging question. We must always remain watchful for just such opportunities. Because prayers are answered in a variety of ways, the answer to our prayers may walk right up to us on two feet (Phil. 4:4).

We also need to be vigilant and watchful for challenges. An old friend once told one of us, "Be grateful for the challenges, for only by facing up to them can you grow." We think God is wonderfully creative in the lessons he teaches us. These lessons are opportunities for personal growth.

We all have weaknesses and blind spots. We all, if we are really honest with ourselves, have to admit to lacking confidence in certain areas of our lives. We all have issues that frighten us. The challenges God sends us afford us opportunities to fill in these gaps and strengthen areas of personal weakness.

We can be assured that the challenges that God gives will never overwhelm us. God's challenges may stretch us, but it is through stretching that we grow. We feel certain that God's purpose is to provide us with all possible opportunities for personal growth. While challenges hold an element of risk (by pushing us into the unknown), we can be confident we always move forward supported by the knowledge that God knows we can succeed. Raise your sights above the heartless laws of the world to Christ.

A practice would not be complete without techniques for confronting the negative. As we have already discussed at length, navigating the confrontations of good and evil is our lot and work as human beings. The first element of our defense is the recognition that evil has no substantive reality, but be aware that it nonetheless exploits our free will. Remember, evil, unlike God, is not required to respect our free will. Therefore evil will always try to trick us in any way that it can to make bad choices. However, our chief strength lies in the fact that the choice is ultimately ours alone; evil can tempt us and lead us in ways that seem very controlling, but evil can never compel us against our will.

The hallmark of a bad choice is that it offers us a brief rush of pleasure followed by a prolonged payback period of intense personal pain that can last for years or perhaps a lifetime. The life that buys into evil moves from brief moments of hot pleasure to prolonged periods of debt repayment in the form of prolonged suffering and decay. A debt to evil must always be repaid with interest!

Lives that fall into temptation and buy into evil are lives that fall into a downward spiral where more and more bad choices are made in a desperate attempt to lessen the personal pain of an individual's previous bad choices. If this downward spiral continues unchecked, it will ultimately lead to a person's destruction, which is often accompanied by tremendous pain and suffering for the people who are closest to him or her.

The only redemption for a person who is heading down such a spiral is for the victim to carefully analyze what is going on in his or her life and accept the reality and inevitability of the pain resulting from bad choices of the past. It may be very hard at the time, but the individual must find the courage to cut his or her losses and make a personal stand for a better future. By mustering all of one's God-given strengths, one can start seeing evil for what it is and start making godly choices for good. Look to the cross and remember the sacrifice that was made there. Begin anew in faith.

Human emotions play a large role in this entire process. It is amazing how people can creatively rationalize their own poor emotional choices. Remember the Flip Wilson[124] character, the Rev, who, when confronted by his deacons over his latest bad choice, always said, "The devil made me do it." The devil never made anyone do anything he or she didn't want to do! However, evil is very clever at offering temptations and ensnaring its victims.

Sometimes it is extremely tempting to seek temporary relief from a difficult situation by releasing our anger at another person. In the moment of such an outburst, the angry person feels really good, but once the anger has been spent, the angry person must find some way to deal with the family member, coworker, or stranger who will not feel very trusting toward him or her in the future.

We have used anger as one example, but those who are temped into embracing any of the survival emotions will act in a way that will play out in a similar scenario. Remember, our choice is made when we display our emotions to others. It is a natural part of life to feel these emotions; this is simply the process of owning our own natures as a part of life on earth. But once the choice is made, if it is a bad one, we must live with the pain of its consequences. If it is a good choice, we will reap the rewards of achieving the inner feelings of goodness and the outward bond of positive connections with our fellow human beings.

We conclude with a final word to all scientists and followers of the one truth of science. It is very important for you to recognize and respect religious people of all paths for their sincere efforts to know better the Creator of our natural world and the Creator's works. We are talking about the very same natural world you have devoted your life to understanding. To all religious people, by all means embrace the scientific knowledge of the natural world. This knowledge is simply the way in which we human beings, with our logical intellects and data-collecting sense organs, come to better understand the miraculous ways of God's creations and draw closer

---

[124] Flip Wilson (1933–1996) was an American comedian and entertainer.

to God. The language of science (empirical data, logic, and mathematics) and the language of religion (poetry, music, art, metaphors, ritual, and myth) are quite different, but the goal of understanding God's creations is the same. The truth that lies at the end of each of these paths is the *same*.

And last of all, to all people who are both scientists and religious, know that what you may think are two different sides of your life are not separate halves at all but one complete whole. The truth can never be subdivided. A path to God and the knowledge of the truth of God's creations can only lead to the same place, and that place is God! *Amen!*

CHAPTER 15

CONCLUSIONS

Our book points the way toward the uniting of the truth of science and of religion. Their long-standing divorce goes back to at least the time of Galileo and his near burning at the stake at the hands of the seventeenth-century inquisition. The unity we are pointing to has generated a number of radical reinterpretations of both the scientific and the religious understanding of the truth. Since science, whose language is measured data, logic, and mathematics, and religion, whose language is poetry, myth, art, and music, have no common language with which to communicate, a way needed to be found to form a basis for a unity that is beyond language. To achieve our goal, we have turned for direction to the field of physics.

Physics has a long and successful tradition of advancing the generality of its knowledge by using a technique called paradox resolution (by reconciling two equally valid but contradicting truths). Examples of successful paradox resolution from the field of physics are Galileo and Newton's development of mechanics, thereby solving the paradox of earth-bound matter vs. celestial matter; Boltzmann's recognition of the arrow of time by solving the paradox of time reversibility, which is allowed by Newton's mechanics; and finally Einstein's development of general relativity as a direct result of his solution to the paradox of two descriptions of mass (gravitational mass vs. inertial mass). Following these examples from physics, we seek to define a paradox that captures the essence of this science vs. religion divorce and search for its solution.

We show that the science of thermodynamics and the science of information theory join forces to prove a principle that every bit of information must contain a minimum amount of energy. A bit is the smallest indivisible amount of information (think of a bit as an atom of information) that can exist. Since all languages are composed of information, any general conclusions related purely to information will be equally true in all languages, including those of science and religion. This means the smallest amount of information relating to either science or religion must be associated with some minimum amount of energy.

Theists are religious people who believe God is present and always with them at a very personal level. For the theist, God truly takes an active interest in how human beings live. Individual hopes, dreams, sufferings, and defeats are all of interest to the God of the theist. If the belief of the theist is to be true, God must have the ability to communicate directly with the theist at any time and any place. For these communications to take place, regardless of their mechanism, God must send a message to the theist, which would require some minimum amount of associated energy.

Without messages from God, theists would have no reason to believe God even exists, let alone know that God is present in their lives. It is widely believed by nearly all religions that God exists apart from the space-time continuum of our material universe. This belief is necessary so God cannot be thought of as being trapped within our finite material universe. This being the case, God's communications to a theist must introduce a certain (nonzero) amount of energy into our universe, in direct violation of the scientific law of conservation of energy.

And here is the paradox: if God is to truly communicate with human beings, then God must violate God's own law of conservation of energy, which holds that the total amount of energy within our universe can never change, even by the slightest amount. If God violates the law of conservation of energy then all of science will fall, but if God does not communicate with human beings, God's very existence and plan would be unknown to human beings, and consequently all religion would become impossible.

The solution we propose to this paradox invokes quantum mechanical measurements as God's method for communicating with human beings and everything and everyone else in our material universe. (Quantum mechanics is science's most fundamental scientific understanding of the subatomic world.) Science regards quantum mechanical measurements as statistically causal (that is, if you were to repeat a quantum measurement a thousand times and average the results, a predictable pattern would begin to emerge). But the measurement is noncausal as far as individual events are concerned. (That is to say, you will get a different answer each time an identical quantum measurement is performed.)

Science regards these individual outcomes of a quantum mechanical measurement as random, unpredictable, and meaningless. We propose to solve the paradox of science and religion by requiring that both science and religion regard each individual quantum measurement as divinely causal. That is, God's own hand reaches down into our material universe and writes God's messages into our material universe without violating the laws of science in any way. By so doing, God's messages are able to fly under the radar of science's conservation laws and remain undetectable from a scientific perspective. Each individual quantum measurement may not convey much information, but bundled together with billions or perhaps trillions of simultaneous divinely causal measurements (in a process we call quantum synchronicity), they add up to God possessing a substantial messaging ability.

It is important to note that there are at least as many human brain cells as there are galaxies within our universe (exceeding 100 billion). Each brain cell is potentially the site of a quantum mechanical measurement as our brain's synapses are switched on and off. By coordinating a large number of these quantum brain measurements, God is able to send God's message directly into the consciousness of an individual human being (and perhaps other living and nonliving things). Based on these observations, we propose a radical reinterpretation of quantum mechanics that recognizes the divine causality of each individual quantum measurement.

In about the year 400 CE in a Middle Eastern country that is now Turkey, a group of the bishops of the new Christian church met to discuss and arrive at an agreement of what was to become a universal set of beliefs for the new church. The results were creeds that today's churches call Trinitarian theology. Trinitarian theology forms the basis of belief for most of today's Christians. Trinitarian theology described God's nature as a unity but with a diversity of three (Father, Son, and Holy Spirit). At a similar meeting of bishops held several years later, it was agreed that the nature of the man Jesus was that Jesus was fully human and fully God all at the same time!

We have compared Trinitarian theology with quantum mechanics and found that each system's framework is functionally identical once proper correspondences have been made. Therefore we conclude that, like quantum mechanics, Trinitarian theology describes the method by which God communicates with us and with our material universe. We see the Trinity as being not so much a description of God's nature as it is a description of the methods used by God to communicate with our material universe. The theologian Karl Barth once said God's being is in God's becoming. Like quantum mechanics is to the truth of science, the Trinity is religious truth's description of how God communicates to us what (in Karl Barth's words) God is becoming. God, speaking through the prophet Isaiah, has already told us, "My thoughts are not your thoughts, and your ways are not my ways" (Isa. 55:8). Therefore we propose to radically reinterpreting the Christian Trinity as (like quantum mechanics is to science) a religious description of how God communicates with human beings and everything and everyone else in our material universe.

Darwin published his *Origin of Species* in the mid nineteenth century. The driving force behind Darwin's theory of evolution is the concept of successful mutations. In the 1940s and 1950s, Schrödinger, Watson, Crick, and Franklin convinced the scientific world that, in terms of microbiology, the essence of life has a quantum mechanical origin. Therefore, the very mutations that drive life's evolutionary change are based on quantum mechanical processes.

By recognizing, as we have, that God's method of communicating originates in quantum measurements, we see how God's communication abilities are uniquely positioned to have a profound effect on the very mutations that drive changes in living things. The religion vs. Darwin's theory of evolution debate has long been the principal battleground between religion and science, and nowhere is it more fiercely fought than in America. However, we are proposing a radical reinterpretation of how the life-changing mutation process (and therefore all of evolution) exists as an act of godly communications to life.

We see evolution as a partnership between God and all living things. God plants mutational seeds that nudge the evolutionary process along in directions that are consistent with God's plan. Thereby all of the scientific aspects of evolution (as described by Darwin) remain in effect because God's mutational seeds must go forth and survive in a new and changing environment if their new (and successful) genetic codes are to become a permanent part of the species' gene pool. From this perspective, evolution is no longer purely a process of hit-or-miss chance. Instead it is led but not directed by God's mutational seeds. God, as the creator of life, is seen to be not so much an architect of life as a fatherly inventor figure who is trying out new ideas to see what will work (and perhaps what will not work) in our world of free will and change.

Human beings have inherited their genetic background from their animal ancestors. Along with our physical traits, these genetic inheritances include the many survival emotions that are needed for each of us to remain alive long enough to pass along our own genetic legacy in the competitive world of evolution. However, as human beings, our highest calling remains the search for God. We are uniquely qualified for this search given our incredible central nervous system that allows God to communicate his ways and plans directly to our bodies.

One of the most important messages we humans receive from God is what we are calling the spirit of virtue. As we confront life's challenges, we all take our own genetic survival emotions too seriously from time to time. All too often we humans fall into the excessive indulgences in emotions such as fear, anger, hate, greed, lust, and pride, to name a few. Sometimes we are tempted into believing in the validity and profound importance of these more negative emotions, tempting us to act on them in ways that may get us into lots of trouble.

A principal way in which God communicates with us (as is necessary for any theist) is to remind us of God's will in the form of a spirit of virtue. The spirit of virtue, if we will let it (this is where free will comes in), will transform us by turning the more negative expression of our survival emotions into their positive twins. Examples are fear being transformed into courage, anger into patience, hate into love, greed into generosity, lust into affection, and pride into humility.

The key to human understanding of this principle is to recognize that our survival emotions are simply a genetic legacy from our evolutionary past whose exercise is always a choice we as individuals must make. God stands ready to help each of us with the transformation of what is emotionally worldly into what is emotionally godly. Perhaps it is the ultimate destiny of the human race to genetically embody (by way of this process of evolution by mutations) God's virtues within the emotional content of our shared genetic legacy. The time may be coming when the fittest among us may be those individuals who are gifted with overflowing compassion, caring, and love. Consider Jesus and the Buddha as examples of what may be humanity's ultimate evolutionary direction.

Ever since people first looked up at the stars, they have asked the following question: "Is there anyone out there like us?" This is not an easy question to answer. Cosmology calls this question Fermi's paradox after Enrico Fermi, who posed this question in a very scientific way back in 1950. We discussed the results of some specific calculations based on the science of communications. From them we concluded that if we were to receive radio transmissions from other planets orbiting other stars, our earth-bound limitations (and the laws of science) prevent us from receiving messages from distances greater than about two hundred light years.

Two hundred light years is actually a very tiny distance when measured in galactic terms and is well short of our best estimates (from the Drake equation) of how far away our nearest technological neighbors might be located. Therefore we must conclude that by using what science knows today, it is very unlikely that we will ever be contacted by intelligent life elsewhere in the universe. Perhaps it is a part of God's plan to keep the various technically advanced cultures around the universe from coming in contact with each other for each to develop and grow in its own unique way.

We discuss how the principle of quantum entanglement (based on Einstein's EPR Gedanken experiment) might theoretically hold the promise of communications over truly cosmic distances. The possibility of using quantum entanglement for communications across cosmic distances is highly speculative, but theoretically it cannot be ruled out.

All our lives are driven by and derive meaning from what we practice. We recommend that all religious people open themselves to scientific knowledge of all kinds, including the latest developments across the full spectrum of scientific understanding. Likewise we recommend that scientists adopt an attitude that their work is first and always focused on discovering and recognizing the awesomeness of God's works. To this end, we think it is very important for both scientists and religious people to make sacrifices at each other's altars. The best offerings we can bring to that altar are our own humility and love. Truth will not be captured or held by any of her creatures. Try to arrest her and she will escape your grasp as did the youth at Jesus's arrest (Mark 14:51) who escaped the gang of temple guards.

But if you love her, she might surprise you with glimpses of her beauty. She surprised an old rabbi who loved her but sought her only in the Torah, the only truth he knew. She met him in the guise of an old donkey driver who questioned him with riddles and opened him to new vistas on the path to her heart (a story from the Kabbalah[125]).

Concerning the truth and the paths in search after truth, consider the path of hypothesis, research, and proof of the scientist on one hand and the path of faith, trust, and enlightenment of the religious person on the other. Truth herself unites, as we have seen, both paths. Curiosity is the searcher's and the researcher's motivation. Intuition is the guiding light and inner voice for both.

The responsibility of both is to God for those people who follow their example and depend on their findings. It appears to be easiest to follow the findings and judgments of science. The scientist goes forward on the basis of facts that have been established and proven by his fellow researchers. The problem is this: facts change, sending the researcher back to the drawing board to find out what happened. We who crave certainty tend to invest the scientist with the ultimate authority to demonstrate truth once and for all until we see him in his laboratory, his mind a question mark. Truth, where are you hiding now? The medical scientist suffers more than others do under this unjust burden. The patient demands a cure. The doctor may be able to affect it or not. All he has to work with are the facts before him and the precedence of similar situations he has seen before.

Faith follows the teacher of truth. The covenant as of old that has been confirmed by that teacher is to ask for what is wanting in your life and your understanding of life. Know and believe that your request is granted. Stay alert to receive what you requested as it reaches your perception and fulfills the practical need that motivated your request. For most people that is a very difficult task. It is like walking a tight rope across the Grand Canyon. There is always that moment when you become aware of the enormity of the task—the moment of doubt. It threatens to dissipate your faith and plunge you into the darkness.

The truth is absolute; however, our perception of truth is tentative. Both science and faith are vexed with uncertainty and doubt. But no effort on the path of truth is ever wasted. Truth patiently waits to be discovered and rediscovered to enlighten our efforts with her blessings.

Truth presents herself to us in principle but also conditionally under varying circumstances. Two major pathways in the search for truth are

---

[125]   Daniel C. Matt, *The Essential Kabbalah: The Heart of Jewish Mysticism* (Edison, NJ: Castle Books, 1997), 138

science and religion. When comparing physics with theology, we found that both disciplines confirm the truth in principle as the evident and undivided truth.

Now as we enter the field of human conduct in our search for the unity of truth, we might find that field the most divisive of all. This is because here we look at ourselves and find that each of us has arrived at the starting point of this search burdened with preconceived notions and fixed ideas. We can confirm as true the fact that we find ourselves on earth as one human species among many living creatures. But how and why do we find ourselves here, and what is our role in relationship toward each other? Those questions result, more often than not, in divided rather than united truth.

The principal truth, from our human perspective, is the objective truth. I say, "I am. I am here."

"That is what you say," says my neighbor. "From where I am, I see you not here, but there. My problem is that from there, smiling at me, you make me feel safe—today. Tomorrow you might frown on me and thus cause me to fear you."

Fear of conflict enters the principle truth of our self-awareness and introduces a modified truth from the subjective perspective. From this perspective, the truth that survives the potential struggle is superior to the truth that succumbs. Among scientists, Charles Darwin clearly observed and described this ubiquitous phenomenon. Success in the struggle for a species' survival falls to the one best fit to live in its environment. The motor of success in adapting a species to its changing environment was first presumed and then confirmed to be the phenomenon of mutation.

We humans have no problem with that observation—that is, until we observe the phenomenon within our own kind. People live in societies and communities in different habitats around the globe. We have been fitting our social and communal lives to our specific environments long before Charles Darwin added his observations to humanity's culture-specific practices.

As an example, Jose Ortega y Gasset,[126] the Spanish sociologist, described the function of ritual in the preservation of peace in human society. Ortega observed that the more sparsely populated an area is, the

---

[126] Jose Ortega y Gasset (1883–1955) was a Spanish sociologist who wrote *The Person and the People.*

more elaborate the greeting ritual. Groups of strangers approaching each other, say, in the Sahara Desert, organize themselves in preparation for the meeting, sending messengers with gifts ahead as soon as they come into each other's view. In progressively denser-populated areas the greeting ritual gets ever less elaborate. One or both hands might be held up as proof the approaching person is not armed.

Fritz writes: I experienced a curious twist of such a situation a few years ago on Hegenberger Road in Oakland, California. As I was filling my vehicle with gas, a man approached me. He opened the flaps of his jacket and said, "I am unarmed. I was just robbed two intersections from here. I am not from here. I'm from Sparks, Nevada. Can you help me out with fifty dollars so I can make it home?" If that was a ritual, it was a twisted, clever use of opportunism and the power of suggestion to induce fear rather than a greeting ritual as described by Ortega y Gasset.

Ortega's description of the greeting's progressive brevity goes to the handshake, to the lifting of the hat in passing (reminiscent of the medieval knight's lifting of the helmet's visor for identification), then to the salute, and the spoken greeting in passing until in our congested city streets, people pass each other wordlessly without acknowledgment. There, says Ortega, the greeting is replaced with the policeman, whose function is to keep the peace (hence the title peace officer).

The truth of that social scientist's observation of the greeting ritual is admirably confirmed in the Bible's ancient story of Jacob's meeting with his brother Esau upon his return to Canaan (Gen. 32-33). Jacob, in fear of his brother Esau's revenge, organized a most elaborate greeting ritual.

Another one of Charles Darwin's scientifically correct observations was misconstrued and is used to this day with harmful results in human society. This is the concept of social Darwinism. This doctrine is used to justify the concentration and manipulation of money, wealth, power, and armed might in the hands of a privileged few in the era of steel and steam. The exploited working people were not convinced that their condition and the sacrifice of their lives and limbs were in any way akin to the evolution of living forms of life in nature, as observed by Darwin. They organized to resist the exploitation and called for justice—an ancient call. Where it was heeded, there was a return to general wellbeing and prosperity. Where the call to justice was ignored, it led to social disintegration, strife, and war. To this day the doctrine of might makes right poisons the minds of many and infests our cities with gang warfare.

The honest truth seeker on the religious path will empathize with the social scientist's dilemma, who tries to communicate an observation only to see the truth of it succumb to falsification and greed. From the seat of truth, the process of establishing its judgment is beyond the range of our thoughts. There is no equivocation at truth's dwelling place: "For my thoughts are not your thoughts, nor are your ways my ways, says the Lord" (Isa. 55:8).

The mandate for human conduct in nature and society differs significantly from that of every one of our fellow creatures: We must not kill our fellow humans—not literally and not in thoughts of resentment. The Bible makes this clear from the outset in the drama of Cain and Abel (Gen. 4:8ff): "Cain attacked his brother and killed him." Everywhere in nature, fratricide is part of the quest for dominance over the herd or tribe. Jane Goodall once observed a chimp getting hold of a metal canister. The racket he produced with it earned him the top spot.

A tomcat will kill a rival male and destroy the rival's offspring and then impregnate the rival's harem with his own seed. This is the observed rule of survival. As we have previously stated, "We are hard-wired for survival. This survival instinct is coded into our DNA helix." In humans that fact is particularly calamitous because of the emotional burden of ever thinking about injury.

The story of Abel and Cain illustrates the truth that we humans differ from our fellow animals precisely in this: the hard-wired survival function in the evolution of the creature species on earth must be overcome in humans if we are to fulfill our function on this beautiful blue planet. We are to nurture life and administer the dominion we have been given with responsibility (Gen.1:28).

The Lord asked Cain, "Where is your brother Abel?" He answered, "I do not know. Am I my brother's keeper?" Yes! We humans are required to overcome the murderer within (who all too often poses as a harmless competitor). No one can claim that is easy. The very Bible's pages are full of descriptions of war and death. It only proves that we humans have a long pilgrimage of pleading for the strength of peace ahead of us. And if we thought the lessons learned in the Hebrew scriptures are tough to follow, Christ, the master of our faith, is quoted in the New Testament with his even more difficult demand: "Love your enemy!" (Matt.5:44) and with the warning, "Those who persecute you will think they do God's will" (John 16:2).

Responsibility is key to the fulfillment of the difficult task of overcoming the violent survival instinct. The question is, to whom or what am I responsible for my conduct? We have two laws we might choose from to answer that call: morality and ethics.

The adherent to the moral law adjusts his conduct to the norms of acceptable behavior. These laws are written in people's constitutions, codes, and traditions and operate within the taboos of a given community and culture. Fellow humans are elected or appointed to oversee the adherence to the laws. The problem is that humans can be deceived and laws can be corrupted.

Ethics, by contrast, are the rules of responsibility to the ever-present spirit of truth—the spirit who calls our consciences from within and without: Where is your sister? Where is your brother? How well do you administer the assignment you have on earth?

Lord, have mercy.
Lord, guide us with your eternal truth.

To contact the authors, please send emails to:

allensweet@aasweetphd.com

dcnfran@aol.com

stellaandfritz@aol.com

INDEX

Page numbers followed by *n* indicate note numbers.

Natural selection
versus the Bible debate, 122
as a partnership between God
and all living things, 208-9
sacrifice and, 179-83
Exodus, 174-76
Experience, 190-99
change and, 190-96
faith and, 196-99
opportunities for, 201
Extraterrestrial life, 156-66
EPR paradox, 164-66
evidence of, 157-61
isolation and, 163-64

F

Faith
determinism and, 70-75
experience and, 196-99
God and, 75
humor and, 199
truth and, 196-99, 211
Fermi, Enrico, 53n57, 156n109
Feynman, Richard, 13n24, 128
Fitzroy, Captain, 115
Foucault, Michel, 70n67
Fourier, Jean Baptiste Joseph,
23n40, 24
Fourier analysis, 23-24
Franklin, Rosalind, 109n92, 208
Friedman, Alexander, 2, 2n8
Frisch, Otto, 37, n49

G

Galilei, Galileo, 4n15, 40
Gauss, Carl Friedrich, 70n71

Gedanken experiments, 39-40,
164, 210
double-slit, 32, 33, 34-38
in Godly communications, 63-67
single- and double-slit, 27-28
single-slit, 28-29, 30
Genesis 1, 2, 11-12, 175
Genesis 2, 190-91
Genetics. *See* DNA
"Ghost function," 63
Gnostics, 81, 197
heresy, 81n77
God. *See also* Religion
actions of, 147-48
challenges given by, 202
Christian theologians and, 80-83
communication with, 56-57,
102-3
control of events, 124
description of, 76-103
diversity of, 100-101
eternal nature of, 99
existence of, 137-38
faces of, 143-44
faith and, 75
flow of information and, 67
Gedanken experiments and, 63-67
human quest for, 136-38
incarnation as Jesus, 101
as influence of outcomes, 74
inspired mutations and, 130
language confusion and, 93-94
laws of, 144-47
mathematics and, 77
memories of, 142

CPSIA information can be obtained at www.ICGtesting.com
Printed in the USA
BVOW032342201212

308818BV00002B/173/P